QUASI-EXPERIMENTAL
APPROACHES

CONTRIBUTORS

HAL J. BOHNER is a graduate student in the Department of Civil Engineering at Northwestern University.

DONALD T. CAMPBELL is Professor of Psychology at Northwestern University.

JAMES A. CAPORASO is Associate Professor of Political Science at Northwestern University.

RAYMOND DUVALL is Assistant Professor of Political Science at Yale University.

RICHARD D. SHINGLES is Assistant Professor of Political Science at Virginia Polytechnic Institute and State University.

MICHAEL STOHL is Assistant Professor of Political Science at Purdue University.

LESLIE L. ROOS, JR., is Associate Professor of Administrative Studies at the University of Manitoba.

NORALOU ROOS is Associate Professor of Administrative Studies at the University of Manitoba.

MARY WELFLING is Assistant Professor of Political Science at Virginia Polytechnic Institute and State University.

QUASI-EXPERIMENTAL APPROACHES

Testing Theory and Evaluating Policy

Edited by
JAMES A. CAPORASO
and LESLIE L. ROOS, JR.

NORTHWESTERN UNIVERSITY PRESS

EVANSTON   1973

CONTENTS

FIGURES

TABLES

PREFACE

For the editors this book represents an effort to bring together work which they and their contributors have pondered, discussed, and read in numerous drafts. It also is an opportunity to assemble in one volume a representative portion of a growing body of fugitive articles on quasi-experimental analysis. Quasi-experimental analysis is a hybrid, reflecting elements of laboratory experimentation as well as features of more naturalistic modes of inquiry. In the various articles, the authors set forth the assumptions, logic, and methodology of quasi-experiments.

There are several features of the book which should aid the reader. A glossary of terms which may be unfamiliar to the reader is provided. Because of cross-referencing and duplication, the chapter references have been combined and are presented as a bibliography at the back of the book.

Financial support for the book came from a couple of sources. James Caporaso received aid from the Office of Research Coordination of Northwestern University. Leslie Roos's work was supported by the International Development Research Center of Indiana University and by Resources for the Future.

Where a book owes so much to intramural stimulation and classroom discussion, it becomes difficult to identify, let alone acknowledge, all those who have helped in one way or another. However, we cannot fail to extend our thanks to Donald Campbell of the Psychology Department of Northwestern University for his inspiration, stimulation, and criticism. Though he is by no means in agreement with us on all points, quasi-experimentation would not enjoy the status it has today if it were not for his pioneering efforts.

Ralph Carlson of Northwestern University Press worked with us closely during the early stages of the book's progress. We thank him for his advice and encouragement. We would also like to extend a major debt of gratitude to Leo Hazlewood of Florida State University and Robert Mahoney of Northwestern University, both of whom read and carefully commented on the entire manuscript.

In addition, Ted Gurr, Kenneth Janda, and Herman Weil of North-western University and John Gillespie of Indiana University gave helpful comments on selected portions of the manuscript. Finally, we would like to extend our thanks to several research assistants who aided at various stages in the writing and editing of the book: Mel Frost, Leslie Rogers, and Kathy Sarantos. In conformity with time-honored custom, we acknowledge our debt to these people while accepting all responsibility for errors of fact and interpretation.

INTRODUCTION

From time to time, social scientists might profitably ask themselves why their particular discipline has not developed more rapidly, why theory-building has proved so difficult. The editors of this volume feel that one reason for comparatively slow progress has been a reliance upon weak research designs, designs which lead to results susceptible to a wide variety of causal interpretations. Thus, the purpose of this volume is to discuss, and illustrate through examples, stronger designs—those which fall under the rubric of quasi-experimentation.

The quasi-experimental perspective emphasizes design; the more powerful analytical approaches—such as causal modeling and factor analysis—can, and often should, be used in concert with quasi-experimental designs. Whenever possible, plausible hypotheses are examined by data generated with an attention to physical controls. There is an effort to minimize reliance upon purely statistical controls and advanced data analysis for repairing deficiencies in design.

The rationale for this emphasis on design is provided by the various threats to internal and external validity which characterize social research. These threats, listed by Campbell in the article reprinted in this volume, generally cannot be alleviated by extensive data analysis. It is primarily by better design that such threats as history, maturation, statistical regression, and experimental mortality can be handled. Better design helps eliminate rival causal interpretations, increasing one's confidence in the empirical foundations of social science.

Although a number of specific quasi-experimental techniques are suggested in this volume, the quasi-experimental approach implies more than learning a new technique or two. This approach is characterized by an effort to use the logic of experimentation in situations which are not truly experimental; in such situations the investigator cannot randomly assign individuals to groups and cannot control the administration of the treatment or stimulus. Quasi-experimentation relies on naturally occurring situations to provide data for the researcher. Information gleaned from archives and records of various types, from situations where the researcher has access but not control, is often used.

An effort to think in experimental terms, to examine plausible rival hypotheses, provides an intellectual structure for the researcher. Given the quality of the research underlying most social science knowledge, such a structure is most useful. At the same time, we have been aware of the theoretical biases implicit in every research approach. The introductory essays have attempted to outline the assumptions underlying each of the designs. Specific research examples are provided in each of the substantive sections.

It should be emphasized that the quasi-experimental perspective is not fixed. The creation of new designs—and uncovering new threats to validity—depend on the imagination of the researcher. The investigator may apply the logic of experimentation to some very diverse siutations. Thus, particular events may affect a time series and be analyzed according to the interrupted time-series design, but the research setting is very different from the classic experimental situation. In other kinds of situations, individuals might be rotated and assigned to groups; some might receive one treatment and some another. But questions of random assignment may remain unanswered, or only partially answered. This case, also, may be treated as a quasi-experiment. As more researchers become aware of the potential of quasi-experimentation, new designs will undoubtedly be developed to fit a variety of new situations.

There are a number of unresolved questions. Several of the individual articles in this volume point out that sampling and measurement errors have not received adequate attention in most quasi-experimental research. Not enough effort has been directed toward highlighting the points of agreement and disagreement with the various causal modeling approaches. Future work along these lines seems both desirable and likely.

Another unresolved problem concerns the links between the different quasi-experimental techniques discussed here. Many variables can be conceptualized in different ways; the same data— or overlapping data sets—can be used in various quasi-experiments. Although an effort has been made to spell out the assumptions underlying each technique, further work on translating among techniques is called for. Guidelines indicating when particular techniques are suitable should be developed further.

Despite the problems, the possibilities are many. The quasi-experimental perspective suggests new ways of looking at both methodological and substantive problems, new ways to address both theoretical and policy issues. This latter concern is reflected in the book's sections on testing theory and evaluating policy. Most important, this book is intended to serve as a stimulus, to convince other investigators of the utility of considering quasi-experimental designs in their present and future research.

PART ONE

APPROACHES TO RESEARCH DESIGN AND DATA ANALYSIS

QUASI-EXPERIMENTAL APPROACHES TO SOCIAL SCIENCE
Perspectives and Problems

JAMES A. CAPORASO

During the last two decades a variety of methodological approaches has surfaced, developed, and become partially integrated into the mainstream of social research. Indeed, such a "multimethodological" emphasis may be seen as one of the distinctive characteristics of the post-behavioral era (Haas and Becker 1970), in contrast to the weighty role of survey research in the early development of behavioralism. Today, simulation, participant observation, aggregate analysis, small-group experimentation, and survey research are all part of an ongoing tradition, and the value of each is not subject to serious controversy.

Nor should it be thought that this methodological diversification results from a fascination with novelty and an indiscriminate desire to tinker with different tools. The cumulation of different methods has proceeded with admirable sensitivity for the development of theory. The early naïve emphasis on alternative methodologies as a series of neutral but independent efforts to assess the same reality (a position essential to convergent operationalism) gave way to the notion of "method-theory" packages, in which each method was seen as entailing a theoretical bias. Hovland (1959) has demonstrated how survey research and experimental work produce conflicting results in studies of attitudinal change. His efforts have been directed toward "reconciling" the differences rather than toward presenting a rationale for their convergence. Similarly Scheuch (1966) has explicated the underlying macro-social bias of aggregate data analysis, while Galtung (1967) has illuminated our conceptual understanding of the ties between survey research and attitudes. Although it has not been fully recognized, I think these developments have pushed toward a complementary rather than a convergent operationalism. Emphasis has been on the differences in access, perspective, and theoretical focus associated with each different method. Thus, rather than an independent set of methodological perspectives converging and replicating the same theory from a different angle, the angle itself has partially entailed a different portion of the theory. What the diversity of perspectives has done is to tap different facets of a problem and increase our understanding of how these fit together.

Yet there are few theories (or developed perspectives) of data collection that can compare to theories of measurement such as those found in Coomb's *Theory of Data* (1964) or Torgerson's *Theory and Methods of Scaling* (1958) or to theories of data analysis such as the elegant explanations of multiple regression models (Draper and Smith 1967) or factor analytic models (Harman 1967; Rummel 1971a).

Several inroads have been made, but they constitute the barest beginning. Dexter's suggestions on interactive (and, by implication, "reactive") interviewing and his discussion of this concept within a framework of attitude theory set the discussion of the problem within the appropriate context (1970). Bruyn's (1966) sensitive treatment of participant observation and his successful attempts to link this methodology to a *verstehen* epistemology solidify our appreciation of this approach. Denzin's efforts to couple the insights of symbolic interactionism to a series of methodological positions are taking shape as a promising approach (1970, 1971), and have already forced the search for a new unit of observation and analysis, new sampling techniques, and new cuts at the age-old problems of reliability and validity.

THE EXPERIMENTAL METHOD IN THE SOCIAL SCIENCES

One methodological approach which has received scant attention in the social sciences is the experiment, despite the fact that its rationale and technology have been carefully refined in physics and psychology. LaPonce (1970:9) notes that in the 1969 issues of the *American Political Science Review* the most popular data-acquisition technique was the questionnaire survey. The experiment, by contrast, accounted for only 3.2 per cent of the empirical-quantitative articles. Similarly, Mueller remarks that

> the value of experimental techniques has only recently impressed itself upon political scientists. Through the 1950s, political science literature was nearly barren of studies based upon experimental methods, but since then interest has increased greatly. (1969:89)

There are some important exceptions to the reluctance to experiment. There is, after all, a firmly established game theory orientation in the social sciences. In addition, simulation, pioneered in the 1950s by Harold Guetzkow and his associates, has now blossomed into a research school of its own. Both are versions of the experimental method, since both entail some manipulation of independent variables as well as nonstatistical controls.

In this chapter I wish to set forth some simple, but essential, components of experimental design. Many advanced experimental designs are very complicated, but they are nevertheless elaborations of a few simple ideas. These simple components will serve as a basis for introducing the idea of a quasi-experimental design. As I will try to explain, the movement from laboratory experiments to quasi-experiments is essentially the movement from procedurally based to functionally based experiments.

An experiment is essentially a process of carefully controlled observation and inference. It falls into the same category as other observational procedures such as participant observation, content analysis, documentary analysis, and survey research. In its laboratory version the experiment's distinguishing features are its active involvement in creating the data one observes and the rigor with which the observation process is controlled.

Let us try to construct an "ideal" experimental process and to isolate its most essential parts. At least six concepts are necessary: dependent variable, independent variable, subjects, confounding variables, manipulation, and control. A *dependent variable* is a variable which we are interested in explaining. It may be the learning abilities of grade school children, the fluctuations of economic development in Ghana, or the differences in civil liberties between England, the Federal Republic of Germany, and the U.S. In all cases we wish to explain why a variable behaves in a certain way. An *independent variable* is a variable used to explain one that is dependent. It is usually thought of as being causally prior to the dependent variable and subject to the experimenter's *manipulation;* i.e., the experimenter can intentionally create variations in it. If, on repeated attempts to manipulate the independent variable, substantial changes occur in the dependent variable, and if the most plausible *confounding variables* are eliminated as rival explanations, one may conclude that the independent variable had an effect. When elimination of confounding variables is achieved, *control* is said to characterize the experimental situation. Finally, *subjects* are the units on which the observations are made. In psychological experiments these are often individuals, though in the biological sciences nonhuman subjects are utilized and in the social sciences larger groups may be observed as entities. In the articles included in this volume, the units on which observations are made include nation-states, individuals, formal organizations, government agencies, and a supranational organization.

Despite the fact that the experimental process is a model for drawing disciplined conclusions from observations, there are many reasons for the lack of experimental studies in the social sciences. One has to do with the disciplinary origins of the experimental method. The experiment emerged and matured into a refined form of controlled observation in the disciplines of physics, chemistry, psychology, and education. An equally important reason was the belief that experimentation was a rigorous technique that was just not relevant for the social sciences. According to this view, experimental analysis demanded isolated laboratory settings, full ability to manipulate the independent variable, and the ability to randomly assign subjects to exposure to the independent variable (and to keep some nonexposed). In the majority of cases the social researcher is interested in situations where he has no ability to manipulate the independent variable (most often it has already occurred), where individuals have been self-selected into groups (not randomly assigned), and where there are a great number of simultaneous events which offer rival interpretations to the presumed causal efficacy of the independent variable being studied.

The assumption on which the papers in this book are written is that the experimental method has much broader application than its laboratory version suggests. This implies no quarrel with the position that characteristics such as the ability to manipulate independent variables and to assign subjects randomly are difficult methodological requirements to fulfill in the world in which the social scientist operates. What is important is not ability to manipulate and to assign randomly, but the ends these procedures serve. Ability to manipulate is merely a way of assuring that there is variation in the independent variable, while ability to randomly assign subjects to experimental and control groups is one of the most refined ways of controlling for the operation of interfering variables. Each is only a specialized technique (and, especially in the case of randomization, a very efficient one). Both are the specialized expressions of the laboratory experiment.

The problem then becomes one of providing the proper translation rules to get the social scientist out of the laboratory and into the "real world," while retaining some of the strong inference characteristic of the laboratory setting. In the following paragraphs I will try to clarify the meaning of the term "experiment," explore the general rationale for experimenting, and finally present a quasi-experimental design suitable for the evaluation of typical problems in social research.

DEFINITION OF EXPERIMENT

As LaPonce has noted, there is disagreement between strict and loose constructionists in the definition of experiment (1970:2). To the loose constructionists (Abraham Kaplan, Marx Wartofsky, and Richard Synder), experimentation means "first of all producing data which would have otherwise remained dormant, . . . poking at nature, waking it up and watching." For the strict constructionist, on the other hand, an experiment implies full control over the stimulus creating the data and at least partial control over other variables potentially affecting the causal interpretation of the manipulated stimulus.

Julian Stanley provides a definition of experimentation in terms of the procedure of random assignment.

> I shall define a true, variable-manipulating, controlled comparative experiment as an investigation in which experimental units are assigned in a simple-random or restrictively random manner to at least some of the experimental combinations. (1967:5)

Skinner, on the other hand, defines the process of control (the general principle of which random assignment is a special case) as the core of experimental analysis (1969:81). Wiggins sees both manipulation and control as crucial (1968), the former providing for interpretable variation, the latter acting to neutralize the noise-filled ecology within which we interpret this variation.

What these definitions have in common is that they are formulated in terms of procedures rather than functions. They thus deprive generality to the simplicity of experimental logic. Fisher once noted that experimentation is "only experience carefully planned in advance" (quoted in Kaplan 1964:147). Experimentation is carefully controlled observation carried out under conditions designed to separate the influence of various variables as well as to minimize errors of observation.

> What experiment can do is to minimize the errors of observation that are inseparable from casual encounters, or at any rate from unplanned ones. The experimenter knows what he is letting himself in for, and is in a position to judge soberly whether (and in what respects) the game was worth the candle. Where errors are not prevented, experiment facilitates their detection and correction. (Kaplan 1964:147)

Experimentation then is a version of observation, distinguished from other versions in its corrigibility and the nature of its design, whose purpose is to detect various casual influences.

THE LOGIC OF EXPERIMENTATION

The root meaning of experiment is close to that of experience. The root of experience is *experimentia,* meaning trial, proof, or knowledge. The root of experiment is *experimentum,* meaning trial or test. The notion of trial appears to be central, from the most disciplined trials carried out under carefully controlled conditions to the most undisciplined probes. Trials may differ in terms of their vicariousness, with "experience" generally reserved for direct physical encounters ("experience is the best teacher") while vision and thought are more symbolic (hence vicarious) applications of the same principle. Many have argued that these differences are not fundamental (e.g., Skinner 1969 and Campbell 1960a), and that the "alternative formulation of the behavior of Man thinking is glimpsed as one of the more distant reaches of an experimental analysis" (Skinner 1969:83). In the latter case the direct physical encounter is substituted by a modeling or "rehearsal" of the environment.[1] In either case the notion of trial as attempt, as probe, is important. Sometimes those trials involve "meddling," or injecting man's influence into nature. A laboratory experiment in which the experimenter alters the payoff matrix in a game of strategy is an example of this. Sometimes these trials are bountifully sup-

1. For an elaboration of this point of view, Campbell (1960a).

plied by nature, as in the manifold policies, reforms, natural disasters, social crises, and so forth that occur without the intention of those studying these phenomena. It is true that many withhold the label "experiment" from naturally produced discontinuities, but this reservation reflects not so much any substantial difference in the process of inquiry as an alienation of the scientist from his subject matter. If a scientist provokes a change it is one thing; if "nature" is responsible for a similar change in the "real world" it is something different. The very usage of the term "experiment" reflects a deep observer-observed distinction in scientific thought.

Experimentation as the optimal vehicle of disciplined, controlled observation is the ideal form of inquiry, at least insofar as that inquiry takes place within the context of verification. Where the experimenter not only can schedule the collection of data (i.e., when and on what units the data are collected), but also has some control over the scheduling of independent variables (i.e., who is exposed to the variables and who is not), and where countless potentially confounding variables can be eliminated through the ability to randomly assign subjects to treatment and control groups, a strong inferential basis is provided which is difficult to approach through other modes of observation. However, the social scientist is often deprived of these procedures because he is interested in processes which take him outside the laboratory, such as the impact of a political campaign on voter turnout, the reasons for failure of a five-year plan, or the causes of economic development. To compensate for this, Campbell and Stanley (1963) have assembled a variety of designs labeled quasi-experimental. These designs attempt to approximate the logic of the laboratory (through again not its procedures, hardware, and techniques) through careful exposure of the many sources of error which can weaken our inferences.

The logic includes three components. The first is the observation of some interpretable variation in some independent variable—a variation of a magnitude sufficient to produce a discernible shift in supposed dependent variations. In the laboratory these changes are physically produced. This is the core of what is meant by experimental manipulation. In nonlaboratory research one must look to "natural" or socially given observations to find clear instances of change. Second, one must observe covariation between this change and change in some other variable. In other words, did the change in the independent variable have an effect, and did this effect go beyond what one would expect on the basis of chance?

By "chance," we mean that part of a variable's variation which is not explainable by an independent variable and which is random (i.e., noncorrelated) from observation to observation. Calculating whether or not a nonchance change occurred is not a simple matter. What is chance or not is partly a function of the underlying theoretical model. If, for example, taking the time-series data with which these articles deal, we make the assumption that all observations were sampled from the same population, then, if the time-series variable contains a normal growth or decline component, significant results can be obtained by cutting in anywhere in the series. The normal probability model used to calculate chance regions assumes a normal distribution of observations around the mean, a condition which obviously does not hold in the case of time-series variables with trend. This is true because each successive observation has a greater probability of having a higher (for positive trend) value than the one before it. Thus other classes of error distributions must be used on which to base the calculation of chance regions.

The third component of laboratory logic is that one must attempt to increase his confidence that the observed covariation between these two variables is a nonspurious covariation by examining the most plausible rival hypotheses. This can be accomplished either through research designs which incorporate some variant of "matched groups," whereby units of observation are selected so that they are as similar as possible on all characteristics except the supposed independent variable, or, at the data analysis stage, through the use of statistical controls such as partial correlations.

Experimentation, then, is controlled observation. But from this it should not be supposed that it is no different from other modes of observation. Experimentation is designed to efficiently assess the impact of trials. By providing the illumination for clear interpretation of the impact of a set of variables, it provides an optimal reality-testing system. As Campbell puts it:

> Experimentation, even if nature answers in the subjective language of the experimenter rather than in the pure tones of the Ding an Sich, has a selective efficiency far above natural selection. We can learn the value of rauwolfia for high blood pressure through three thousand years of Dravidian custom selection, but we get with it a lot of irrelevant and ineffective pharmacological lore that it would take another thousand years to weed out. Experimentation can achieve a winnowing of equivalent purity in one decade or less. (1969b:23)

QUASI-EXPERIMENTAL DESIGNS

Among the classical objections to the experimental method are its lack of realism and its inability to produce large effects (i.e., substantial reactions on the part of the dependent variable to variations in the independent variable). In an experimental setting, realism is sometimes difficult to achieve since the participants, aware that they are involved in an experiment, do not respond seriously to the stimulus situation. Perceptions of crisis and threat are likely to be higher for decision-makers participating in an actual war than for decision-makers simulating a war. Because experiments transpire in a contrived and simplified environment, the results obtained usually have limited generalizability (low external validity). One can agree with this without getting into the argument over whether experiments are "real" or not. As Meehl has pointed out (1970), whatever happens is real, but the stimulus field of the lab is not the stimulus field of many other situations to which we wish to transfer our findings. Of course one may include the peculiar features of the laboratory as part of the specifying conditions of the theory so that the theory is said to hold only under these conditions. Transference of the theory to the nonlaboratory setting would pose little danger to the generality of the theory, since the necessary specifying conditions would be absent. One may pursue this line, but only at the cost of making the theory trivial by attaching so many special conditions to it. There may be a law to explain the crisis behavior of Yale sophomores from specified backgrounds, over a certain IQ level, at a certain time of year, being paid $2.00 an hour to take part in the experiment. Yet all this may change if the hourly rate or any one of the other specifying conditions is altered. The second limitation, inability of experiments to produce large effects, stems from the same problem of the contrived setting. This objection may seem to run counter to much of what is proposed as an advantage of experimentation. It is true that the experimenter can manipulate the independent variable not only in terms of presenting it or withholding it but also in terms of varying its strength. But "strength of the stimulus" is not an exclusive function of the stimulus-object (source of the stimulus), as Dewey and Bentley have argued (1949:242), but also of the target of the stimulus. And the absence of an adequate transactional specification for "stimulus" has led to all sorts of embarrassments for environmental determinists who view changes of magnitude of the stimulus-object as an adequate explanatory vehicle.

The development of a series of quasi-experimental designs by Campbell and Stanley (1963) was partially a response to several of the limitations of traditional experimental design. Quasi-experimental designs are rooted in conditions where there is no possibility of manipulating the stimulus and no control through matching and randomization over competing stimuli. Experimental stimuli often occur naturally, many times with no active intervention of the researcher. This helps to cut down on problems of external validity as well as small experimental effects. The price, of course, is the increased problems of control.

I will outline the features of five quasi-experimental designs. (For more extended treatments, the reader is referred to Campbell and Stanley [1963]; Campbell [1968]; French [1965]; Crano, Kenny, and Campbell [1972]; and Pelz and Andrews [1964].) This discussion will not attempt to detail all the threats to the internal and external validity of all these designs, nor will it discuss tests of significance separately for each design. A discussion of the most common threats to internal and external validity and of tests of significance for one of the designs, the interrupted time-series analysis, will follow.

The five quasi-experimental designs to be discussed here are: (1) the nonequivalent control group design; (2) the field experiment; (3) the cross-lag panel correlation design; (4) interrupted time-series design; and (5) control series or multiple time-series design. Although not all of these designs are termed quasi-experimental by Campbell and Stanley (e.g., cross-lag panel designs are included under correlational and ex post facto designs), we put them under the same label since all at least partially fulfill the three prominent characteristics of quasi-experimental designs. They all attempt to approximate or simulate manipulation, to provide controls for confounding variables, and to probe the data for causal dependencies.[2]

NONEQUIVALENT CONTROL GROUP DESIGN

This design is extremely useful in judging the effects of a variable on a group where that group has assembled naturally, i.e., has not been brought together by the experimenter for his own purposes. One may be interested in assessing the differences between two classrooms taught by different methods; the effects of a medical

2. This particular way of summarizing quasi-experimental designs was suggested to me by Leo Hazlewood.

drug on a group of patients, a portion of whom have been non-randomly assigned the drug; or, as in Shingles' article in this book (chapter 7), the effects of involvement in Project Head Start on its members compared to a group of nonequivalent nonmembers. This design is essentially an extension over the one-group pretest-posttest design ($O_1 X O_2$). In addition to the experimental group, which is measured both before and after exposure to the experimental event, there is a control group, which is measured at the same two points in time but which is not exposed to the event. We thus have

$$O_1 \, X \, O_2$$
$$O'_1 \qquad O'_2$$

where the Os refer to observations on the dependent variable and X represents the experimental event. Each now represents a separate group. The advantages of adding the control group lie in the additional base for comparison. The one-group pretest-posttest design, which is represented by $O_1 \, X \, O_2$, has only one comparative base. Inferences that differences between O_2 and O_1 are due to the occurrence of X are plausibly rivaled by other variables occurring jointly with X as well as by trends or cycles in O. For example, if a variable increases or decreases by a fixed amount at each time point over a long series even where X is not present, conclusions about O_2–O_1 differences attributable to X are fallacious. By adding a control group, even though it is not equivalent to the experimental group in all respects, one controls for both of these threats to validity. If other variables and trend effects are operative, they should affect both groups. The logic of this procedure is fairly clear: adding an untreated group makes certain threats to valid inference constant for both groups. What differences we observe between the two groups can therefore more confidently be assigned to other variables, among which, of course, is X.

There are, however, still some threats to internal validity which remain uncontrolled. Since the groups are nonequivalent on an unknown number of variables, it is possible (though not always plausible) that there is some interaction between X and variables specific to the experimental group. One possibility is that there is a "selection-maturation" interaction (Campbell 1968:260), where differential rates of maturation (different trends and cycles) are associated with the distinguishing features of experimental and control groups. Similarly, a threat to external validity is provided by the possibility of interaction between selection bias and X. In this case X indeed would have an effect, but such effects

would be limited to populations sharing the selection characteristics of the experimental group. Adoption of this design also limits
the experimenter to analysis of differences between means rather
than comparison of slopes and growth curves. Finally, it seems that
this design will catch only immediate effects and will not be able
to detect delayed responses.

THE FIELD EXPERIMENT

Field experiments refer to those inquiries which take place in
natural settings where the investigator actively intervenes to
manipulate at least one independent variable.[3] Field experiments
qualify as quasi-experimental designs because they manipulate
variables and search for causal dependencies. They thus satisfy
two out of the three characteristics of quasi-experimental design;
the third characteristic, control, is met in different degrees by
most field experiments.

Field experiments differ from experiments carried out within
the laboratory in that in the latter the "investigator creates a
situation with the exact conditions he wants to have and in which
he controls some, and manipulates other, variables" (Festinger
1965:137).

The essential difference between field and laboratory experiments turns on that vague thing called *setting*. This refers to
nothing less than the total environment in which the experiment
takes place. This total environment must be seen as a complicated
field of stimuli in which each stimulus possesses the potential to
influence the dependent variable. Inside the walls of the laboratory, there is a great reduction in the complexity of this stimulus
field, and this simplicity facilitates the interpretive tasks of the
investigator. In field experiments, however, the process of inquiry
must intrude into settings in which people convene naturally,
i.e., as part of their normal social intercourse. Thus schools, super-
markets, places of work, and summer camps are all potential
settings for field experiments.

Field experiments also differ from "natural experiments," i.e.,
those situations where an experimental interpretation is brought
to bear on an event or process which has already taken place or
which will take place in the future without any manipulation on
the part of the researcher. So while field experiments differ from

3. For a comprehensive discussion of the field experiment with fruitful comparisons between this and laboratory experiments, see French (1965).

laboratory experiments in terms of *setting,* they differ from natural experiments in terms of *manipulation.* As French puts it:

> In the field experiment, the manipulation of the independent variable is not left to nature but is contrived, at least in part, by the experimenter; thus, the design is planned beforehand. (1965:99)

It may also be helpful to compare field experiments with the nonequivalent control group design. In terms of formal representation it need not differ. For example, we can construct a field experiment along the following lines, which is exactly the same (formally) as the nonequivalent control group design employed by Shingles:

$$O_1 \, X \, O_2$$
$$O_1' \quad O_2'$$

The difference, of course, is that in this case X is manipulated by the investigator. If the investigator administers the stimulus abruptly and with sufficient strength, his task of assessing its consequences will be considerably easier than in the kind of natural experiment which Shingles has carried out. Of course, the possibility of *reactive error* is greater in field experiments, especially when manipulation of variables is recognized by the subjects. However, a more compelling reason for not doing field experiments (for Shingles as well as for the rest of us) is that many times one simply has neither the ability nor the desire to manipulate. This becomes obvious when one is studying the effects of crisis on war.

Field experiments are compromises between laboratory experiments and natural experiments, and from this vantage point they offer several attractive features. To some extent, experiments in field settings get around the problem of artificiality associated with the laboratory. Because of the complexity and naturalness of the experimental environment, people are more involved and less aware of their experimental role. For much the same reason, field experiments usually have greater impacts, or effects, on participants.

> The more realistic the research situation, the stronger the variables. . . . Realism simply increases the strength of the variables. It also contributes to external validity, since the more realistic the situation, the more valid are generalizations to other situations likely to be. (Kerlinger 1964:383)

But because field experiments are compromises, they mirror the tensions and dilemmas that exist between research traditions in which the emphasis is put on control and precision and traditions in which nonartificiality and complex relations are favored. The greater the nonartificiality of the research, i.e., the less contrived the setting, the more realistic the experiment will appear to the participant. But realism implies nonsimplification of the setting and noninstitution of conspicuous controls. Where all variables are allowed to freely intermingle, it will be difficult to draw confident conclusions. To the extent that visible manipulation is used in the field experiment, and to the extent that visible controls are instituted, the participant may become more aware of the experimental situation and react in a subject role. Block and Block have pointed out that subjects of middle-class background "structure the experimental situation as one calling for a submissive role in relation to an authority figure" (1952, cited in French 1965:100). As French quite correctly points out, where experimental findings are a function of special role behavior, there is a severe threat to external validity (1965:100).

THE CROSS-LAG PANEL DESIGN

Many times we are interested in the behavior of two variables we suppose to be causally interrelated. In situations of this type, there is a temptation to choose one of two "all or nothing" hypotheses: X causes Y or Y causes X. It is possible that both of these hypotheses are correct in varying degrees. The cross-lag panel design is appropriate for these problems and it has some capacity to distinguish between relative causal impacts. To be more specific, it can discriminate patterns of *preponderant causality* (Crano, Kenny, and Campbell 1972) even where mutual feedback is involved. In the paper by Duvall and Welfling, this design was adopted to deal with one of three theoretical problems they tackled, the relationship between social mobilization and political institutionalization.

The basic logic of the design is as follows:

FIGURE 1.1.   Cross-Lag Panel Design

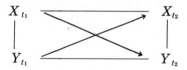

There are two variables measured at two points in time (t_1 and t_2) from which six correlations can be generated. When the synchronous correlations are stable ($rX_{t_1} Y_{t_1}$ and $rX_{t_2} Y_{t_2}$), we are interested in the lagged diagonal correlations (i.e., $X_{t_1} Y_{t_2}$ compared to $Y_{t_1} X_{t_2}$). If the preponderant causal structure is from X to Y, the correlation along the downward slanting diagonal should be greater than the correlation along the upward slanting diagonal.

There are some problems of this design which have been dealt with in various places (Pelz and Andrews 1964; Rozelle and Campbell 1969; and Heise 1970). There are at least three basic problems: the relation of the synchronous correlations to lagged correlations; the assumption of the equality of time lags; and the *necessarily* incomplete theory on which the cross-lag model rests.

There is the assumption that the synchronous correlations are on the average smaller than the lagged correlations along at least one of the diagonals. This assumption is usually only implicitly made, but it seems necessary to preserve a well-defined meaning to causality. If synchronous correlations are consistently larger, two conclusions are possible. Either most of the causal relations between X and Y are instantaneous or the interval of measurement (i.e., the temporal unit of aggregation within which the observations were grouped) is too large, so that it includes most of the causal relations inside each periodic measurement. If this is the case, one can break the measurement interval into smaller units and then lag the variables.

A second problem has to do with the lagging interval, or the time lapse necessary for X to have an effect on Y. In the Crano, Kenny, and Campbell study of the relationship between achievement and intelligence, the lagging interval is two years, from fourth to sixth grade. In the Duvall and Welfling paper (chapter 3), the measurement interval is five years (1955–59, 1960–64, and 1965–69). The lagging interval from t_1 (1955–59) to t_2 (1960–64) is also five years, though this is actually an average figure. Now the assumption, of course, is that the length of the lag is the same for both variables. If it is not, we may obtain the following results where X affects Y at lag one and Y affects X at lag two, but since we do not examine lag two we are never made aware of this.

FIGURE 1.2.   Cross-Lag Design with Differential Time Lags

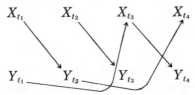

If these results lead to conclusions to the effect that $X \rightarrow Y$ is the preponderant causal pattern, the conclusions will be misleading unless they are accompanied by a conditional statement about lags. It is therefore important to experiment with various lags.

There is a third problem which has for the most part been ignored in the literature. This is the basic incompleteness of the model on which the cross-lag correlations are based. This incompleteness is not by choice of the researcher who has formulated the model; it seems, rather, to be a logical necessity without which the model could not work. To see this, let us imagine the extreme case where X at time one causes all the variation of Y at time two but Y_{t_1} causes none of the variation in X_{t_2}. The application of this model would then be limited to one case. Since Y would have no feedback effect on X, X would be incapable of further variation and, consequently, X could not continue to affect Y. Applications of the model to $r_{t_2 t_3}$, $r_{t_3 t_4}$, and so forth depend on the operation of exogenous variables. Of course, we are dealing with cases where feedback does exist, but the same principle holds in an extended version. If X affects Y strongly but Y only affects a small part of the variation in X, there is an essential incompleteness in our understanding of X. Without taking variables additional to X and Y into account, Y would have a progressively diminishing impact on X and vice versa. Since Y only explains a small part of X (say, 20 per cent), each successive series of X observations will be only partially explained, and if we take this part which is partially explained it must explain a progressively smaller part of Y. This is a positive feedback process which should result in a progressively smaller impact of each variable with each successive time period.

The approach is still useful in helping to discern causal patterns, but it is most appropriate for those situations in which strong mutual feedback is involved so that mutual influences do not "dampen" from one time point to the next. One of two final conditions should also hold. Either the theory should be approximately complete with X and Y or, barring this, the error terms (in this case mostly the error term due to exogenous variables) should be constant from one time point to the next.

THE INTERRUPTED TIME-SERIES DESIGN

This design is appropriate to data distributed over time (we shall refer to such a distribution as a series) and where there is theoretical reason to believe that some event should cause a change in the behavior of the series. Stated in more precise terms, this design involves (1) periodic measurements or observations on some variable at equally spaced points in time, (2) the occurrence of an event somewhere in the series, and (3) the assumption that the event occurs midway between two selected measurement points. Finally, this design involves a critical evaluation of results in light of those hypotheses which pose the greatest threat to the hypothesized relationship. This design may be diagramed as follows:

$$O_1, O_2, \ldots O_m, O_{m+1}, O_{m+2}, \ldots O_n, O_{n+1}, O_{n+2}, \ldots$$

$$\uparrow \qquad\qquad\qquad\qquad \uparrow$$

$$\text{Event 1} \qquad\qquad\qquad \text{Event 2}$$

where O represents periodic observations on some variable over time and where events 1 and 2 represent occurrences thought to have an effect on the behavior of the series.

The key question involved in this design is whether the occurrence of the events in question had an effect or whether the behavior of the series after the events represents an undisturbed continuation of the series from its previous state. This is by no means a simple question and cannot be resolved by a simple visual inspection of plots of the data. Two questions are involved: Did a nonrandom change occur in the vicinity of the experiment? Is this change attributable to the occurrence of the experimental event? The resolution of the first question is the task of tests of significance, used in conjunction with appropriate theoretical models. The second question is more a problem of the validity of interpretation. It thus involves untangling a variety of potentially operative causal factors, and selectively eliminating those which are plausible but false.

Let us direct our attention to the first problem and see what a significant change might look like. Figure 1.3 illustrates a variety of possible patterns a series may assume.

Suppose each of these lines represents the behavior of a dimension of integration over time. X represents the occurrence of some important event, such as the formation of the European Economic Community, the acceptance of a landmark piece of legislation concerning agricultural policy, or the elimination of tariffs. If the European Economic Community is a system in the sense that it is composed of a set of coordinated social structures, then these events should have an impact on other variables. In figure 1.3, lines A, B, and C show no discernible effect, while lines D through G do suggest that changes have occurred, though in different

FIGURE 1.3.   Possible Patterns of Behavior of a Time-Series
Variable (adapted from Campbell and Stanley [1963:38])

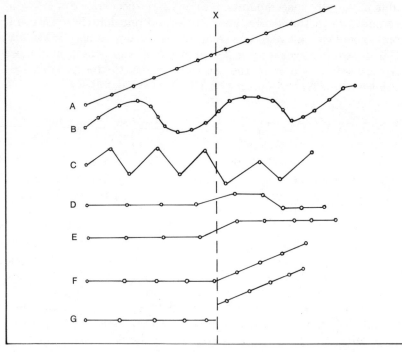

fashions. Let us look at these individually for a moment. Line *A*
indicates no effect of *X*. It is true that the post-*X* values are higher
than the pre-*X* values, but this result is obtained by a simple
extension of pre-*X* values and indicates no distinct effect of *X*. The
line is said to contain a general (i.e., monotonic and, in this case,
linear) trend, or increase over time. This kind of line is commonly
at the basis of most positive growth systems. Line *B* represents a
cyclic pattern where more or less regular crests and troughs are
observed over fairly well-defined time intervals. Although it may
appear that *X* had an effect here, it is clear that the post-*X* pattern
again is predictable by an extension of the properties of the cycle
before the occurrence of *X*. This interpretation becomes clearer
once we have a fuller picture of the cycle. Suppose, however, that
X occurred two observations earlier than it did in the figure, and
suppose further that we only had information on a fragment of
the cycle. The figure would then look like this:

In this figure line *A* is actually a temporal fragment of line *B*. If
we only observed *A* it would appear that *X* forced an authentic

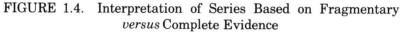

FIGURE 1.4.   Interpretation of Series Based on Fragmentary *versus* Complete Evidence

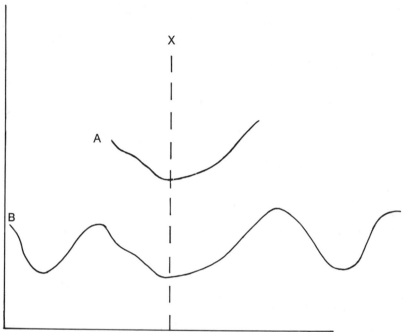

change in the subsequent operation of *A*, when in fact the increase in *A* is a function of the orderly continuation of the cycle. The presence of a cycle is not always so clear cut. It is not uncommon for several cycles to be present in the data simultaneously and to be nested in one another. Similarly, line *C* represents no true effect of *X*, though in this case this is due to random fluctuation in the variable — not to the presence of cycles. Random fluctuation becomes a threat to hypothesized interpretations when the behavior of a variable has a high standard deviation.

The other four patterns (*D–G*) represent "true" changes, though whether these are due to *X* is of course another matter. *D* represents the kind of change usually associated with static equilibrium. A disturbance variable causes a temporary shift in the behavior of the series, after which the series returns to its normal (i.e., previous) behavior. *E* represents a dynamic equilibrium change where the introduction of *X* causes a change in the behavior of the series which becomes more or less permanently absorbed into the conduct of the series. The series reaches an equilibrium but it is at a new level. It is important to note that the form of the series does not change. Its slope has not changed although the level at which the series operates is higher. Line F, on the other hand, reflects a change in slope after X. That is, there is a change in the growth rate of the variable after X. This indicates that the operation of the series is pushed up to a new level and, in addition, is characterized by a new growth function.

This design is employed in this book in the chapters by Caporaso, Duvall and Welfling, and Stohl. It is appropriate for extended series into which a relatively discontinuous event intrudes. When these two things are present, a strong basis for inference is provided even when no control series is present. It should be borne in mind that comparisons may be provided by additional time cuts as well as by the addition of novel units.

THE CONTROL SERIES DESIGN

This design is basically an extension of the interrupted time-series design. It is formed by adding a series which is parallel (i.e., for the same time points, for the same measurement interval, but for another unit) to the original. For example, Campbell and Ross (1968), in their attempt to assess the impact of Connecticut Governor Ribicoff's crackdown on speed violators, used a pool of neighboring states (New Jersey, New York, Rhode Island, and Massachusetts) to provide a comparative baseline. The experimental hypothesis was that Ribicoff's December, 1955, announcement that speeding would result in a thirty-day suspension for the first offense would result in a decline of traffic fatalities. Perhaps equally plausible causes of such a decline were changes in the weather, particularly the severity of winters, and safety features associated with the new 1956 automobiles. By adding the additional comparisons from adjacent states, these rival hypotheses were neutralized.

The formal structure of this design is as follows:

$$O_1 \quad O_2 \quad O_3 \quad O_4 \quad X \quad O_n$$
$$O_1' \quad O_2' \quad O_3' \quad O_4' \quad \quad O_n'$$

As with the single interrupted time-series analysis, there are periodic measurements for one variable, the occurrence of an abrupt event (X), and a post-X series. In addition, there are periodic measurements for the same variable on another unit, where X does not occur. It is generally true that the superiority of this design over the single series design is its ability to control for history. However, this claimed superiority should be examined with great scrutiny in light of the particular substantive problem. With the identical control series design, one design may control for history while the other may not.

As an illustration, let us suppose that two classes in the same high school are part of an experiment to assess whether showing a series of movies on race relations affects racial attitudes. One class is shown the movie and the other is not. The movies are shown daily for a two-week period, and at various times during this period tests are administered to see if attitudes have changed. Of course, any other events occurring during this two-week period relevant to racial feelings pose an alternative explanation. The control series design should be resistant to this threat because attitudes should be equally affected in both groups. Similarly, if one is examining the frequency of traffic fatalities in Connecticut and wants to control for weather conditions and safety features of autos, a control series design is adequate.

What should be noticed is the following: the control of extraneous variables is not completely a formal property of the design. It is an induction based on an implicit hypothesis that the variables being controlled are common to both units. To put it more precisely, the general principle is that the rival hypotheses must operate without discrimination on both experimental and control groups. This criterion is likely to be satisfied when the units under examination are small, nonautonomous, and externally dependent on the same sources, so that to the extent these units are affected from the outside, they are commonly affected. This is likely to be the case with adjacent states (such as Connecticut, Massachusetts, Rhode Island, New Jersey, New York) under a common weather system and automobile distribution system. However, should the units be large, noncontiguous, autonomous, and "internally driven," the effects of history are likely to be more unit-specific and therefore uncontrolled. "Largeness" cuts down on diffusion, thus reducing the number of confounding variables which spread to both units. Autonomy and "internal drive" are much the same and to the extent they hold the system will be uniquely affected by historical forces within it.

Many nation-states are large, sociopolitical units with a great deal of autonomy. Problems of developing control series at this level are difficult. They are much more severe at the level of the regional and supranational organization. For example, most regional associations are very large, are usually separated by great distances from one another, and are tied into different global networks. Take the European Economic Community (EEC), which is the subject of one of the interrupted time-series analyses. Its external ties with the United States and the eighteen associated African states are strong. On the other hand, the European Free Trade Association (EFTA) has stronger ties to the Commonwealth. To the extent that each regional organization is affected by its particular external ties or by its separate internal affairs, the results will be specific to the unit. In situations where this is expected, the development of temporal comparisons for a single series is urged.

The previous discussion was concerned with the question, "Did a change occur?" Until a confident answer is given to this question, there is no point in moving on to further analysis, which concerns itself with such questions as these: "What caused the change?" "What variables contributed, and which are spuriously related?" The major substantive explanation of changes, when it is established that they have occurred, is the quasi-experimental variable in question. Since hypothesis testing is more than establishing positive associations and involves creative attempts to falsify hypotheses through a careful exposure of the researcher's hypothesis to competing plausible interpretations, it may be useful to assemble a checklist of trouble spots. The following list is adapted from Campbell and Stanley (1963) with some modifications of our own.

1. *Random Instability.* A change found after a quasi-experimental event is due to random fluctuations in the data. Any distribution has a random error component in it due to observational errors of a nonsystematic nature, sampling inaccuracies, and perhaps also small, nonsystematic changes in the thing being studied. *Corrective:* Tests of significance are designed to inform us whether or not an observed value could have occurred by chance.

2. *Trends and Other Periodicities.* A periodicity will be defined as any repeatable pattern in the series. A trend will be defined as a special kind of periodicity, one characterized by some average increase or decrease, i.e., some mean shift, in the behavior of the variable. Trends and periodicities are what Campbell means by "maturation." They are processes dependent upon the passage of time. *Corrective:* Solutions to the problem of trend (assuming that there is a theoretical basis for wanting to be rid of it) include trend-removal techniques (Box and Tiao 1965; Smoker 1969), adjustments in the interpretation of significance tests (Sween and Campbell 1965), and corrections in the number

of observations to take account of autocorrelation (Quenouille 1952).

3. *Other Substantive Variables.* These include all other events and/or processes which may appear coterminously with X or between X and its presumed effects, and which rival X as the explanation of the observed changes. *Corrective:* The only correctives here are good theory and imaginative use of control groups and matched groups. If, for example, dissolution of trade barriers is viewed as an alternative explanation of subsequent integration, one could select his units such that they were matched (i.e., similar) on trade barriers but different on the hypothesized variable.

4. *Scoring Procedures.* It is possible that changes observed after X may be due to changes in the way the variable is recorded after X, rather than to the effect of X itself. This is a prosaic matter but one which can greatly affect one's results. *Corrective:* If recording changes are made and one recognizes them, one can usually adjust for them (e.g., by adding or subtracting an appropriate constant).

5. *Irrelevant Indicators.* While the change we observe may be a true change (i.e., nonrandom) and indeed may be a response to X, it nevertheless may represent a response irrelevant to the concept with which we are dealing. For example, we may be interested in the impact of certain events on economic integration but our indicators (imports and exports) may be invalid indicators. We know that every indicator and index contains a good deal of variance that is independent of the concept in which we are interested. It may be that the change represents a response on the part of this component. *Corrective:* A strategy of multiple operationism in which a variety of indicators are pooled to form a composite index may be used. Hopefully this results in the cumulation of "true" concept variation and the canceling out of unwanted variance.

6. *Regression Effect.* This refers to the tendency for pretest and posttest scores to converge (regress) toward the mean when the observations prior to the treatment (i.e., experimental event) are chosen because of their extreme values. Campbell and Stanley (1963:10) point out that this is a danger in the evaluation of remedial experiments where subjects are selected because they do extremely poorly on a particular test. This is also a problem in the evaluation of policies, crises (such as the EEC agricultural decisions and

the EEC crisis in Caporaso's article and the wars in Stohl's paper—chapters 4 and 5), and other social phenomena which occur because of extreme values on other variables. Where these extreme values exist, there is a good chance that posttreatment values will lie closer to the mean even if the event had no effect.

TESTS OF SIGNIFICANCE

It is presumably the function of tests of significance to estimate the probability that an observed point or distribution could have occurred by chance. Earlier in this chapter chance was equated with random error, i.e., error which showed no pattern from one observation to the next. The distribution of almost any variable contains some random error. The question to which a test of significance is addressed is whether or not an observed change exceeds the limits of what is expected on the basis of chance fluctuations.

In our case we are specifically interested in whether some characteristic (e.g., the mean, the slope of a line, the value of a particular point, or the value of an intercept of a line) of a distribution of points previous to a quasi-experimental event is significantly different from that characteristic in the distribution after the occurrence of the event. Thistlethwaite and Campbell (1960), Campbell and Stanley (1963), and Glass, Tiao, and Maguire (1971) point out some special problems in the application of tests of significance to time-series data. For example, a simple t test assessing the difference between pre-X and post-X means would produce significant results if trend were represented in the series. Figure 1.5 illustrates this. The line from X to Y' probably has a mean significantly higher than the line from Y to X. However, as Thistlethwaite and Campbell observe, this significance ignores the fact that there is a general regression of both pre- and post-X distributions and that this obtains independently of the quasi-experimental event. In cases such as this, where there is a high probability that the error terms are not normally distributed around the regression line, it makes little sense to calculate chance regions. Techniques have been developed (Lord 1958; Cronbach and Furby 1970) which first remove the trend of the values in a series and then express "residualized" scores in terms of their departures from a general regression line. If the adjacent points of the residuals are also dependent, estimates of autocorrelation must be made separately for pre- and post-X observations. This should be done separately,

FIGURE 1.5. Tests of Significance Where Trend Is Involved

since, if the introduction of X does have a strong effect, post-X observations will show high autocorrelation. Presumably, then, these residual scores could be utilized as the transformed data for appropriate tests of significance.

It is important to notice that we are not saying there is no difference between the pre- and post-X distributions or that nothing has occurred to cause this difference. What we are saying is that the change in question is not peculiar to X but is a general property of the series. It must thus be explained by reference to the operation of more long-term, smoothly operating values. To put it another way, one must tie the language of inference and tests of significance to the underlying theoretical model.

We will utilize four tests of significance here. Each of these tests is based on a calculation of the difference between expected and observed (or expected and expected, in the case of the double-Mood test) values of points or distributions, where expected values are based on some property of the regression line.

SINGLE-MOOD TEST

The single-Mood test is a t test appropriate for assessment of the first value after the occurrence of an event from a theoretical value predicted by an extrapolation from a linear fit of pre-X values. As pointed out elsewhere (Caporaso and Pelowski 1971), it is a simple line-fitting technique based on the least-squares criterion where the regression estimate of the pre-X data is used to predict the first observation after X. This predicted value is then compared with the first observed value. The t statistic yields a value indicating the probability that the observed value could have occurred simply by extrapolating the line. The single-Mood test is suited for distributions of the following type:

FIGURE 1.6   Single-Mood Test

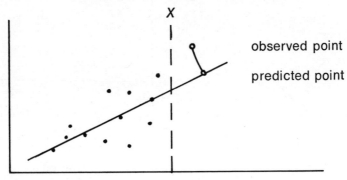

DOUBLE-MOOD TEST

This test simply extends the logic of the single-Mood test. It involves both a pre-X and a post-X linear fit and a comparison of the predictions by these two estimates of a hypothetical value lying midway between the last pre-X and the first post-X point (Sween and Campbell 1965:6). This test is appropriate for the assessment of intercept differences as well as slope changes.

WALKER-LEV TEST ONE

This statistic evaluates the hypothesis that a common slope fits both pre- and post-X data. This condition (i.e., a common slope) may hold even if the occurrence of an event causes a change in the mean level at which the series operates. For instance, a five-year plan may result in a shift in productivity to higher levels without affecting the rate of economic growth.

WALKER-LEV TEST THREE

This test yields an F-statistic which tests the null hypothesis that a common regression line fits both pre- and post-X distributions. Separate regression estimates are calculated for both sets of data. These are subsequently compared to see if they could have been drawn from the same population.

The assumptions on which the statistical model is based are in

FIGURE 1.7   Double-Mood Test as Applied to Intercept
Difference and Slope Difference

Intercept Difference

Slope Difference

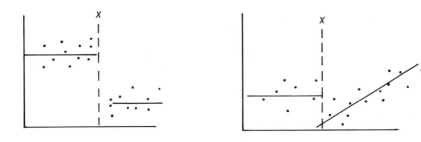

some cases not fully met.[4] This may seem inexcusable in light of
the fact that others have attempted to introduce alternate statistical models. Box and Tiao (1965), Glass (1968), Glass, Tiao, and
Maguire (1971), and Box (1967) have all had a hand in introducing
a statistical approach based on assumptions more appropriate to
most time-series data. For example, the model suggested by Box
and Tiao more realistically assumes that time-series observations
are autocorrelated. Therefore the model makes provision for the
estimation of a parameter to account for this. Their model also
makes no assumption of constancy or uniformity of effect throughout the post-X series. This seems intuitively more reasonable. One
expects that some wearing off or attenuation will take place. Another parameter is therefore estimated to weight more heavily
those observations closest to the event in question.

Although the Box-Tiao model has many attractive qualities,
there are still too many technical difficulties and unmet assumptions to justify immediate implementation for our time-series
data. One of these unmet assumptions has to do with the absence
of trend, which the Box-Tiao model demands. Most of the dependent variables of interest in this volume possess some form of
trend. Still, the application of this statistical model remains one
of the most exciting and potentially fruitful avenues for future
research.

Quasi-experimental designs offer some unique advantages
when compared to orthodox experimental design and ex post facto
research. Perhaps these advantages can be illustrated in terms of

4. This does not mean that violations of the assumptions fail to be taken into
account. For example, estimates of autocorrelation are taken in order to include
consideration of the nonindependence of successive observations.

the effect/error ratio, the amount of variation in the dependent variable due to variation in the independent variable divided by the amount of error. Here I use the term "error" generically, to refer to both variable error (errors of observation, measurement, sampling) and equation error (error due to omitting important variables from the equation).[5] All designs seeking to demonstrate relationships aim for a high effect/error ratio. A large effect makes for interpretable variation, which is needed to conclude that something significant (and substantial) has taken place. A small error term minimizes the effect of disturbing influences and provides a stable and controlled background which serves to throw the effect into relief.

One of the traditional objections to both experimental design and time-series analysis is the low effect (small variation) engendered. In experiments this is ostensibly because of the artificiality of the experimental setting, while in time-series analysis it is because the variable of interest varies less over time than through space. Conversely, this is one of the arguments in favor of comparative research — that by using countries (rather than time points for a single country) as observations one can maximize variation in the independent variable. Practically speaking, comparative designs provide greater discontinuity, more interpretable variation, and less equivocal effects. In time-series designs, there usually is less effect but more closure to compensate. Comparative designs attempt to increase the strength in the numerator but thereby unleash a host of other uncontrolled variables, creating problems in the denominator. Seldom is a matching or "most similar systems" design used, in which the countries are implicitly matched on a number of potentially confounding variables (Przeworski and Teune 1970). Longitudinal designs operate through the denominator, attempting to hold more potentially confounding variables constant. Quasi-experimental designs soften the impact of both of these problems. First, by selecting natural experiments, they deal with an independent variable which takes on the full range of variance. Quasi-experiments occur not in contrived, artificial settings but in the ongoing world of social processes. Second, quasi-experiments focus on events which by definition are

5. Variable error would be zero if we had a perfectly measured variable (true score = observed score) with no instability (high reliability). Equation error would be zero if all the important variables were included in the equation (theoretical completeness) and the relations among variables could be formulated deterministically with no stochastic or error term.

more discrete than many other continuous processes, such as socio-economic development, urbanization, or transnational integration. Events are sometimes contrasted with variables, but this is misleading. An event is a variable which occurs infrequently and which has neater boundaries (origin and termination points) than continuous variables. Thus they are characterized by less border problems than smoother variables. They are easier to locate in time and space and therefore easier to assess in terms of their impacts on other variables. In some of the following papers an event is merely a discontinous portion of a longer distribution. For example, in the paper by Caporaso on the European Community, the agricultural package deal is a condensed version of decision-making in the agricultural sector. Decision-making occurs in this sector on a continuous basis, but the package deals represent an intensified or telescoped version. By focusing on these arbitrarily defined bursts of variance, one can hope to increase the effects in the effect/error ratio.

LIMITATIONS OF THE QUASI-EXPERIMENTAL APPROACH

In this section I will discuss some of the limitations of the quasi-experimental method with two purposes in mind. First, such a discussion will hopefully foster improvement and refinement of the method. Second, I hope it will prevent application of the method to problems for which it is not suited. Some of the limitations are inherent in the method and are not subject to change. These are also bound up closely with the advantages of the design. The first limitation concerning the focus on abrupt events is of this type. Other limitations are problems related to the state of knowledge, and hopefully these can be overcome.

A complete list of the limitations of this approach would have to include the following: (1) its focus on abrupt events; (2) the lack of measures of strength of association; (3) difficulties in the estimation of parameters; (4) its alleged atheoretical leanings; (5) its nontransactional theoretical bias; (6) the difficulty of handling multiple influences simultaneously; and (7) its restriction to time-series data. I will not discuss the estimation problems here since these problems are common to all time-series data. Suffice it to say that most estimators do not yield best, linear, unbiased estimates (BLUESS). Most estimates are unbiased (i.e., the mean value of the estimates is equal to the population value) but not best (i.e., having minimum variance). Neither will I discuss the problems of multiple influences simultaneously, except to say that advance here is dependent on making some progress in deriving measures of strength of impact of events. Until one can evaluate the strength of influence of independent variables, it is impossible to critically assess multiple impacts and to discriminate relative influence.

ABRUPT EVENTS

It is clear that a quasi-experimental approach focuses on abrupt, discontinuous events. The experimental event, if it is not a discrete occurrence, will almost certainly be a sharp change representing a radical departure from previous behavior of the variable. By contrast, the long-term behavior of the variable is used as the background against which to assess this discontinuity. This long-term behavior may itself be described by a general trend. Sometimes these long-term trends are very important, as in the case of the relationship between growing specialization of society and the emergence of voluntary groups (Durkheim 1933) or the relationship between economic and political development. In spite of the importance of some trends, quasi-experimental analysis looks for discontinuities in such variables and examines the impact of these discontinuities on other variables. These impacts may include a discontinuity in a second variable, a shift in its trend, or a shift in the mean level. The above processes are phenomena with long-term secular trends. To understand these one must have a longer temporal range and must acquire a feeling for how these processes develop.

To evaluate relationships between long-term trends, one must utilize other techniques, such as cross-variable time-series analysis and autoregressive models.

Conceptually, then, in the interrupted time-series analysis, a "true change" is one that changes the function in the series, or one at least that is not interpretable in terms of the series' past function. Change will therefore occur under the following conditions: (1) when there is a discontinuous shift away from the trend line due to the impact of a substantive variable; (2) if there is a change in the rate of change (slope) or mean level of the variable (intercept); or (3) if there is an error exceeding the limits of the probability model.

LACK OF MEASURE OF STRENGTH OF ASSOCIATION

The techniques for evaluating whether or not an event had an effect are limited to visual inspection and tests of significance.[6] Thus while we possess a measure of the improbability of a certain effect we have little cue as to how strong the relationship is. One could look at the magnitude of the intercept change or slope change and obtain a rough estimate. However, to my knowledge, the statistics for this kind of evaluation are not well developed. This seems to be more than a temporary limitation, however. There is a logical obstacle presented by the fact that we are dealing with nonpaired, unequal observations in the dependent and the independent variable. Measures of association require that one observation for a supposed independent variable be paired with an observation on a dependent variable. These pairs may be distributed over time or space. For example, we may be interested in relating violence to the level of systemic frustration, and we may do this by looking at both of these variables for a set of countries or for a single country over a set of time points.

ATHEORETICAL NATURE OF QUASI-EXPERIMENTS

In reading Pierre Duhem's discussion of the experiment in physical theory, I was struck by the fact that in physics the questions that became experiments were suggested by theoretical puzzles.

> A physicist disputes a certain law; he calls into doubt a certain theoretical point. How will he justify these doubts? How will he demonstrate the inaccuracy of the law? From the proposition under indictment he will derive the prediction of an experimental fact; he will bring into existence the conditions under which this fact should be produced; if the predicted fact is not produced, the proposition which served as the basis of the prediction will be irremediably condemned. (1962:184)

The experiment is suggested by a question about theory, and the content of the experiment is nothing more than careful observation on matters whose outcome will falsify or confirm the point in question. Thus physical experiments perform a corrective function for theory by modifying, refining, adding, and sometimes even overthrowing.

---

6. This point was independently brought to my attention by Frank Hoole of Indiana University and Kenneth Janda of Northwestern University.

This seems to be very different from what is done here. In quasi-experiments one simply asks whether X had an effect. Did a certain policy bring desired consequences? Did the crisis in the European Community cause a slowing of integration in various parts of the community system? Such questions are limited to what is observable. If policy X did not have certain consequences, that is all there is to it. No implications from this are drawn in terms of an overarching theory. These generalizations seem to be no more than simple summaries of what has been recorded; they do not extend an inductive base to cases not yet observed. For this reason these findings might add to our factual knowledge (which is already voluminous) but not to a cumulative body of theory.

There is a great deal of merit in this line of criticism. Yet I do not think there is anything inherent in quasi-experimentation which necessitates this. The problem is that we so often treat X as a discrete event and do not say what it is an instance of. But then, if X is just X, and not the special occurrence of a more general variable, nothing we discover about X can be transferred to other manifestations of the variable. The most that such an ad hoc approach could lead to would be a patchwork of findings with no coherent framework.

I see no reason why this criticism applies only to the experimental method. Similarly I see no reason why it cannot be overcome. One must simply ask what the causal event in question was an instance of. Take, for example, the first agricultural package deal in agriculture. What does this represent? One could obviously (but nontrivially) say that this is a special case of decision-making in the agricultural sector. This would of course say nothing about the presumed impact of decision-making in other sectors. However, in the analysis of the European Community, decision-making in agriculture is further conceptualized as a special case of decision-making in a functionally undifferentiated sector. In addition, the articles by Duvall and Welfling, Stohl, and Shingles are explicitly tied in to important theoretical questions.

NONTRANSACTIONAL NATURE OF QUASI-EXPERIMENTATION

The last thirty years have seen the convergence of knowledge around general systems theory, cybernetics, and the concept of transaction.[7] This convergence of orientations has resulted in a reinterpretation of many methodological concepts, such as stimu-

7. For a good treatment of some of these developments, see Buckley (1968).

lus-response, cause-effect, subject-environment, and so forth, which were based on the strict separation of actor and environment. This reorientation has created some additional problems for causal thinking. Causal inference is based on the assumption that a causal event occurs independently of the effect (dependent variable). This means that an independent variable is equally likely to occur irrespective of the values of the dependent variable. If this is not the case, the "dependent" variable is influencing the "causal" variable. Under laboratory conditions this independence can be fulfilled through experimental manipulation of the "when" of exposure. In almost all of the quasi-experiments examined here, the event examined may be at least partly a function of the dependent variable. For example, when Abraham Ribicoff introduced the law instituting a crackdown on speeders (Campbell 1968), the law itself was a function of prior (high) fatalities due to car accidents. Social systems are cybernetic systems. Most remedial or goal-oriented activity is therefore likely to occur when a system variable takes on a high value. Failure to take this into account will result in misleading conclusions about the causal relationships involved.

POSSIBLE USES OF EXPERIMENTAL METHOD

The preceding constitutes an impressive list of objections to quasi-experimental analysis. Yet there are many reasons to use this approach. I would like to outline some of them here.

First, the experimental method is useful in helping to ascertain facts that otherwise might not be uncovered. This is one of the points made by Boring (1954) and emphasized by Kaplan:

> Much of the forethought that goes into scientific observation is directed toward making accessible what otherwise could not be seen, or if seen, would not be noticed. Special care is taken to insure that the scientist will be able to see what he is looking for if it is there to be seen. (1964:127)

If experimentation is carefully controlled observation, it follows that factual inquiries should turn up more accurate reports. "Did a significant change occur after event X?" "Was a particular program in experimental education successful?" Experimental analysis can provide answers to causal questions as well as to factual ones. It does this by forcing us to think at the operational level of directly observable behavior and by instituting controls to aid this observational process.

Second, a quasi-experimental analysis may help us to build up a body of low-level empirical generalizations. While this may be some distance removed from the broad-gauge, abstract, deductive theory advocated in philosophy of science books, it has some merit of its own. It puts one in close touch with the data and makes one constantly translate one's theory into concrete observations. Hopefully, empirical generalizations found at this level may dovetail into higher-level theory and strengthen its supportive base.

A third use of quasi-experimental analysis is to test relationships. The researcher may have already constructed a model or some loose hypotheses and may want to use the experiment to check out these ideas. When the number of variables is few, and when complicated interactions are presumed not to exist, this approach may be the most efficient. When the number of variables is large, and when variables are related to one another in complicated ways, perhaps other approaches (such as path analysis) may be more useful.

A corollary advantage should be mentioned at this point. Under certain conditions it is nearly impossible to test relationships except in a quasi-experimental way. A good example of this is provided by the situation in which the dependent variable has a distribution of observations, but there is only one observation on the independent variable. This one observation is the experimental event itself. Technically, two observations are present—one in which the event occurs and one in which it does not occur. For example, the impact of a war on a dependent variable, as illustrated by Stohl's article, could not be tested by any of the traditional measures of association. One could utilize the point biserial correlation if there were a distribution on a dichotomous variable, such as sex, and an interval-scaled variable, such as foreign policy attitudes. But this is more than a problem of relating a dichotomous and a nondichotomous variable. It is a problem of relating a variable on which there is a distribution of cases to one which has two implicit observations. To make the point as generally as possible, the interrupted time-series design enables one to test hypotheses for which there is an unequal number of observations on dependent and independent variables.

Finally, the quasi-experimental approach is advocated as a complement—not a substitute—to the deductive approach. To make this argument we need to diverge for a moment to talk about the relation between experimental laws and theories. The distinction usually made is that the experimental laws contain only terms that refer directly to observables, while theoretical statements are more abstract and contain at least some terms that are not operationally definable. This is not a mere problem of "directness" of observables, as Carnap has pointed out (1956); it involves some fundamental differences of opinion concerning the empirical status of theoretical terms. One position is that not all theoretical terms are translatable, even through a series of intervening operations, into an observation language.

> The rules connecting the two languages [theoretical and observable] (which we shall call rules of correspondence) can give only a partial interpretation for the theoretical language. (Carnap 1956:39)

Once this distinction is admitted, the implications for experimental work become obvious: the level of confirmation of experimental hypotheses must be seen to be at least partially independent of the changing theories in which they are embedded and which in turn may be used to explain them (Hesse 1967:405). If this were not the case, and if the meaning of experimental laws were exhausted by the theories that entailed them, there would be no excuse for not fitting experimental findings into a larger deductive framework. This could only have the effect of increasing the implicative strength of experimental findings by providing support or falsifying evidence for a larger number of abstract statements. However, because rules of correspondence do not provide a comprehensive and isomorphic set of translation rules, experimental findings may survive shifts of theory. As Nagel notes:

> Accordingly, even when an experimental law is explained by a given theory and is thus incorporated into the framework of the latter's ideas, . . . two characteristics continue to hold for the law. It retains a meaning that can be formulated independently of the theory; and it is based on observational evidence that may enable the law to survive the eventual demise of the theory. (1961:86)

The position that experimental findings are compatible with a range of different theories and that theoretical terms usually only provide a partial empirical interpretation presents the temptation of two dangerous extremes. The first is that of extreme operationalism. This reaction could result from reasoning that if the meaning of experimental terms is not supplied by theory, it can only be supplied by the operations conducted in the experiment. The second extreme is the opposite: that all experimental meanings are supplied by a theory which has unequivocal rules of correspondence to experimental operations. The danger of the former position is that it is likely to invite a multitude of incommensurate findings, a mosaic of detail, rather than a cohesive framework of laws. The dangers of the latter approach are that it might deprive potential bodies of theory of a firm supportive base of inductive generalizations and that it overlooks even the possibility that a theory may come into existence in this manner. I suggest a murkier, more moderate position in which theories are not rejected in toto when an experiment provides negative evidence and in which experimental findings are not rejected should the covering theory change. In view of our imperfect knowledge of the languages that coordinate broad theoretical terms to directly observable behavior, this ambiguous path seems to be the preferable one.

> Theoretical experiments are not the only ones by which science advances, even theoretical science; and though such experiments may be the most important ones considered singly, their contribution may well be matched or even outweighed by the cumulative effects of other types. Not every experiment need be designed with an eye to the Nobel prize. The belief that every soldier carries a marshal's baton in his knapsack may heighten morale, but it is not to be forgotten that every marshal needs an army of soldiers behind him. (Kaplan 1964:153)

2

PANELS, ROTATION, AND EVENTS

LESLIE L. ROOS, JR.

In social science much theory-building and policy research must proceed from situations where the independent variable is "socially given" (not under experimental control). Quasi-experimental designs are particularly relevant in such situations. These designs attempt to introduce the logic of experimentation where control over the scheduling of experimental stimuli (the "when" and "to whom" of exposure and the ability to randomize exposures [Campbell and Stanley 1963]) is lacking. In quasi-experiments much effort is directed toward identifying which variable or variables are independent and which dependent. Given the complexity of social phenomena, establishing the preponderant pattern of causation represents a reasonable goal for quasi-experimental research.

This goal may not always be attained. A number of threats to internal validity—history, maturation, testing, selection, experimental mortality, and so forth—are common to quasi-experiments. Two of the most important are:

1. History—events, other than the ("experimental") treatment, occurring between time 1 (time of first observation) and time 2 (time of second observation) and thus producing alternate explanations of effects.
2. Maturation—processes within the respondents or observed social units producing changes as a function of the passage of time per se, such as growth, fatigue, secular trends, etc. (Campbell and Stanley 1963).

Threats to validity characteristic of quasi-experimental designs using longer time series have been treated extensively elsewhere.

How the logic of quasi-experimentation might be applied to data from just two or three points in time has been less often discussed.

Although a quasi-experimental treatment of data from longer time series is certainly useful for policy analysis (Caldwell and Roos 1971), short time-series data may be particularly relevant in this connection. Head Start programs and school integration are two issue areas in which the concerns of policy makers for relatively quick information about the programs' effects led to major cross-sectional data collection and analysis (Williams and Evans 1969; Coleman et al. 1966; Mosteller and Moynihan 1972). But problems concerned with making valid inferences from cross-sectional data are well known; there are almost always a number of plausible alternative hypotheses which cannot be eliminated (see Campbell and Stanley 1963). These problems have led to increasing interest in various designs involving the collection of data from several points in time. Thus, with regard to school integration and busing, the passage of several years has permitted better analysis of a critical policy problem (Armor 1972, 1973; Pettigrew et al. 1973).

It is important to stress that in these quasi-experimental designs the researcher is imposing an intellectual structure upon ongoing events. Each design incorporates certain assumptions about the real world which can be tested, but only partially. This essay will discuss these assumptions and illustrate how — by changing the underlying assumptions — different designs might be (and often should be) used with the same body of data.

SHORT TIME SERIES

Short time-series data have been generated both by researchers working at the macro-level and by those studying micro-level phenomena. This methodology is growing in popularity because collecting data from at least two points in time often allows the investigator to consider time-precedence which, as Crano, Kenny, and Campbell (1972) stress, is a highly useful rule of causal inference. When a change in one variable consistently precedes a change in another, several possibilities should be considered:

1. The change in variable A is a cause (possibly only one of many) of the change in variable B.
2. The change in variable B is a cause of the change in variable A. In a chain of occurrences — A_1, B_1, A_2, B_2, A_3, B_3, etc. — it is difficult to say whether the changes in A preceded the changes in B, or vice versa.

3. The changes in both variable *A* and variable *B* are the effects of some more general cause(s).

Ideally, a research design should be a guide in collecting data so as to facilitate examination of the various rival hypotheses outlined above. Criteria to help in sorting out these different hypotheses are necessary. Equally important, the researcher should be aware of alternative approaches to the problem being considered and take into account the assumptions underlying each design. The quasi-experimental designs discussed in this essay will confront these issues within the limitations of data collected from just a few points in time.

COMPARING DESIGNS

Although other research has applied cross-lagged panel analysis to aggregate data (Duvall and Welfling, chapter 3 of this volume), this essay focuses on survey data – data collected from individuals in social settings. Those factors particularly relevant for micro-level studies are emphasized. Different systems vary markedly along dimensions having real consequences for research designs. Often individuals are linked to clearly defined social situations, such as their jobs or other roles. Quantities and types of roles may increase or decrease as organizations expand or contract. Moreover, cases analyzed by social scientists are often characterized by significant events altering the relationships between individuals and their settings; e.g., bureaucrats are reassigned and legislators are changed by elections.

Although role and setting are somewhat interconnected, they can be separated for analytical purposes. A limited definition of role is adopted here; it is defined as the formal position which an individual occupies. In order to separate role from setting, the informal aspects of role are not considered here. Setting is defined as the place where the individual occupies the role.

A common research problem for students of organizations and elites involves distinguishing the effects of variables associated with individuals, social role, and social setting. Some strategies for using individual mobility among roles and settings to separate these effects are developed in this essay. The emphasis is upon short time-series data – the type generated in many field research situations. The discussion is selective; some designs which are relatively rare, such as the regression-discontinuity design (Campbell, chapter 6 of this volume), are not treated in this essay.

Various research designs, each particularly suited for a situation with certain characteristics, are suggested below.

Type of Design	*Design Characteristics*
Panel Design	A quasi-experimental design based on relationships among two or more variables measured at two or more time points. At the micro-level it is particularly useful when individuals stay in the same setting. The design assumes the system is closed and assumes little turnover. It is suitable for a range of research purposes.
Rotation Design	A quasi-experimental design best suited for situations where individuals are rotated among different settings, but keep the same roles. It is also possible when individuals are rotated among different roles in the same setting. For research purposes, this design assumes that the system is closed. It is appropriate for many bureaucracies and for business.
Pretest-Posttest Design	A quasi-experimental design based on changes in one or more variables before and after a particular event. Various controls are possible. There is also an experimental version of this design. The system is assumed closed, except for the one event.
Replacement Design	A group of research designs to deal with problems of selection and drop-out. Good designs try to consider individuals' recruitment and release within a common framework. It is appropriate for moderate or high turnover situations and for those where the number of roles (positions) is in flux. It seems relevant for some legislatures.

The characteristics of these designs are summarized in table 2.1. In this summary individuals and their settings are treated, providing a convenient way to compare the logic of the different designs.

Setting is important because an individual occupies a particular role in a given setting. If the setting of individual role occupancy is changed, he may be unable to operate at a suitable level. For example, a district manager who moves from Chicago to Houston may face unanticipated problems which affect his performance. In analogous fashion, an individual moving to a different role in the same setting—a manager promoted in the Chicago office—may also encounter difficulties. If the focus is upon the role itself, the concept of replacement may be appropriate; successive individuals

TABLE 2.1

INDIVIDUALS AND SETTINGS IN DIFFERENT RESEARCH DESIGNS

t_1 First Wave	Intervening Event	t_2 Second Wave
	Panel Design	
I_1-S_1	Possible "change" in individuals and/or settings (causal locus of change may be inside system; testing for more general cause)	$I_2 (= I_1)$ in S_2 (may or may not $= S_1$)
	Rotation Design	
I_1-S_1 I_1'-S_1'	Rotation of individuals among settings (causal locus of rotation outside system; locus of change may be inside system)	$I_2 (= I_1)$-$S_2 (= S_1')$ $I_2' (= I_1')$-$S_2' (= S_1)$
	Pretest-Posttest Design	
I_1-S_1	Possible "change" in individual resulting from an event; many controls possible (causal locus of change outside system)	$I_2 (= I_1)$ in S_2 (may or may not $= S_1$; change in setting may be the event)
	Replacement Design	
I_1-S_1	Replacement of one individual by another (causal locus of change not specified)	$I_2 (\neq I_1)$-$S_2 (= S_1)$ (individual is replaced)

Role is assumed to be held constant in all designs.

occupying a particular position in a given setting are considered. For some purposes, individuals may cease to be of interest when they leave this position or setting.

These designs differ in their assumptions about the system being studied. Two of the designs—the panel and rotation—emphasize variables internal to the system as determinants of behavior. These designs might be said to have a "closed system" perspective. Typically in the pretest-posttest design a single external event is considered; one particular "treatment" is seen as affecting one or more dependent variables. For replacement designs the system is assumed to be open; transactions with the environment and changes in this environment are likely to be important influences upon micro-level variables.

The value of a variable may change for one or for a number of reasons. Each of these designs tends to emphasize a single type of reason rather than several reasons combined. The term "causal locus of change" is used in table 2.1 to illustrate this. Variables internal to the system being studied are suggested as the causal factors in panel and rotation designs. One or more events outside the system are seen as important in the pretest-posttest design; in such designs events are characteristically conceptualized as variables. In replacement designs the change in individual units may be responsible for change in variables. The reasons for this switching of units—changes in the population being studied—must be explored, but such reasons are not specified in the design.

Events also can be conceptualized in several different ways. They may affect the shuffling of individual units among various settings; such movement has obvious implications for selection of an appropriate design. But events have other consequences for social research: they can substantively affect the variables with which the analyst is dealing by (1) changing the level of one or more variables, or (2) changing the relationships among variables.

Different research designs treat events in various ways; for example, in the panel and rotation designs, events as sources of change in variables are omitted. If an event can be "reconceptualized" as a variable which changes between time periods, it can be brought into the design as a possible causal factor. Thus, as discussed later in this paper, the event "reapportionment of a state legislature" between t_1 and t_2 could be integrated into a panel design by introducing a "degree of legislative representativeness" variable which took on different values at t_1 and t_2. The event has been "internalized" into the design.

The panel design is particularly suitable for the study of systems characterized by little turnover and little change in the number of positions. The match between data and system must be assessed, since in this design the system being studied is assumed to be closed. With a great deal of recruitment or release, individuals remaining in the same position at several points in time may be of little relevance for system functioning. From a more methodological perspective, the study may have little external validity because of changes in the population. The new (untested) individuals entering into the system may differ markedly from those respondents studied over time. Moreover, as Blalock (1967) has pointed out, if individuals who migrate differ along several dimensions from those who stay, inferences concerning the relationships among variables will be affected.

The rotation design has one great advantage. When the rotation process is close to being random, this design approaches a true experimental design. But without knowledge that randomization was actually employed in the rotation process, inferences about causation must be weaker than with a true experiment. When the researcher works from the data alone, he can only eliminate different measured variables as predictors of assignment. Even if no measured variables are significantly correlated with the setting of assignment at the next time period, unmeasured variables might have affected rotation in systematic fashion.

Events can also be treated as possible sources of change. The pretest-posttest design and the interrupted time-series design (see chapter 1) both assume exogenous events, sharp cutting points, and the ability to divide a time series into "before" and "after" points (or segments). Various statistical tests can be used to test whether the two points (or lines) differ significantly. Thus, in these designs one attempts to evaluate the impact of a particular event. Since individuals generally cannot be randomly assigned to treatment and control groups, various threats to validity must be approached by the use of one or more nonequivalent control groups, whose initial characteristics differ from those of the treatment group (the group experiencing the given event).

Finally, replacement designs can be thought of as a group of designs focusing upon new entrants into the population or institution being studied. As an individual leaves one position for another, or as a new position is created, it is desirable to collect data from the individual recruited for this position. In growth situations, as well as in those with a fair amount of turnover, replacement designs may be integrated with panel or reassignment designs to help maintain some of the advantages of each.

LIMITATIONS OF SHORT TIME-SERIES DATA

Some of the issues with regard to short time-series data are:
1. the direction of causality,
2. the treatment of time,
3. measurement error,
4. "turnover" and nonresponse,
5. changes in measures of central tendency and dispersion over time, and
6. changes in the stability of variables over time.

A common question for short time-series designs concerns the number of time points at which data should be collected (the number of waves). The more time points and the greater the number of indicators, the easier it is to deal with the above issues. The minimum case involves data from two variables at two points in time. With such information a number of plausible hypotheses may be eliminated, but, as Blalock (1970) and others have indicated, several competing hypotheses are likely to remain.

Research using few time points should focus upon preponderant patterns of causation. Hypotheses like "a change in variable A causes a change in variable B" can be investigated. But relatively complex causal interrelationships are difficult to trace using just two or three time points. If a change in variable B feeds back to influence variable A, this will be lost in an analysis emphasizing asymmetrical, one-way causation. Thus, when just two or three points are used, measured relationships among variables may not provide an adequate picture of social processes (Brunner and Liepelt 1970). Models implying reciprocal causation pose problems of data collection, measurement, and estimation which are treated in detail elsewhere (Blalock 1970; Land 1971).

The use of short time-series data also depends on plausible (yet arbitrary) assumptions about time lags. In particular, for this approach one assumes that the interval between cause and effect is short enough to be noted by data collection at the given time points. On the other hand, the period of causal lag cannot be so short as to be instantaneous. "In particular, it is assumed that the period of causal lag is greater than the time required to measure one sampling unit" on both variables (Heise 1970:5). With "instantaneous" causality, any kind of lagged analysis would be meaningless. This is not to say that variables A_n and B_n (variables A and B measured at the same point in time, point n) need be uncorrelated. As Heise (1970:6) has pointed out, correlations between such variables would be likely because of past operation of the causal system. The interval between observations must be long enough to pick up changes in variables, yet if the pattern of observations is markedly out of phase with social causation, the researcher's conceptions of the system being studied may be in error.

Remedies for this problem are being investigated. With one-way causation and highly stable variables, Pelz and Lew (1970) have suggested that two-variable causation can be untangled when the measurement lag is substantially longer than the causal interval. If data from a number of time points were available, several lagged correlations could be used, varying the length of time lag. This would help in disentangling likely causal relations, but it is clearly no panacea. Some guidelines for this kind of work have been produced by simulations, where a change in one variable leads to a change in another variable. Formulas and assumptions can be varied, producing sets of data with known properties. Such questions as the match between the causal interval and the measurement interval can be fruitfully investigated with this technique (Heise 1970; Pelz and Lew 1970).

Additional points should be made with regard to measurement. Considerable benefits accrue by using measures on which previous research has been done. The selection of measures which load on different factors can aid in rejecting a hypothesis which poses a significant threat to causal interpretation, the hypothesis that one variable — rather than two or more — is being measured (Brewer, Campbell, and Crano 1970).

As has been pointed out elsewhere (Duncan 1969b), differences in the reliability of measures used in causal inference can complicate model-building. If the reliability of the measures is known, corrections for unreliability or attenuation can be made, producing corrected correlations. Working with corrected correlations "should give a closer approximation to the theoretically underlying relationships than would manipulations on the uncorrected data" (Duncan 1969b:83), although problems of correlated measurement errors are not accounted for by these corrections.

A central problem with short time-series data is "separating out real change from measurement reliability" (Blalock 1970:1103). Selecting measures for which reliability has been estimated can help assess a major rival hypothesis to causal inference — namely, that the measures were unreliable. But reliability can not be handled in an absolute fashion. New populations and new circumstances may change the relationships among measures independent of any intended treatment effects; items having substantive meanings and intercorrelations at one time period may not at another. For example, Pool, Abelson, and Popkin report that

> the question "How well do you like (Senator Joseph) McCarthy?" did not correlate with the identical question asked at different times. . . . The question had changed its entire meaning at the time that the United States Senate censured him. (1964:39)

When a priori estimates of test-retest reliability are unavailable, several strategies are possible. The more traditional approach is to use test-retest correlations in conjunction with efforts to minimize temporal instability (actual change in the measure between time periods). Such efforts characteristically involve the use of relatively short time periods and individuals whose circumstances have remained constant between t_1 and t_2. Another approach focuses upon properties of the respondent in addition to those of the measure. Converse (1970) has emphasized that a subset of the population may respond randomly to various questions; measures of reliability will be deflated by including such individuals in the calculations. Dividing the respondents according to their personal histories may also be helpful in estimating reliability. In ongoing research, subdividing a panel study's sample according to the amount of change in the respondents' objective circumstances may permit rough estimates of the lower limit of reliability.

A newer approach treated in detail elsewhere is path analysis (Boudon 1968; Land 1969). Path coefficients are standardized regression coefficients which may be thought of as the ratio of two standard deviations. As noted by Blalock (1970:1103), "in the denominator is the actual standard deviation in the dependent variable, whereas in the numerator we have that portion of the standard deviation in the dependent variable that can be attributed to the particular independent variable under consideration, with the remaining causes controlled."[1] Path analytic procedures have been developed "that enable one to handle random measurement errors whenever there are multiple measures of each variable" (Blalock 1970:1103). Given certain assumptions,

---

1. A path coefficient is a number which measures the fraction of the standard deviation of the dependent variable (with the appropriate sign) for which the designated variable is *directly* responsible in the sense of the fraction which would be found if this factor varies to the same extent as in the observed data while all other variables (measured and unmeasured) are constant. (Taken, with minor changes, from Land 1969:8–9.)

such procedures make it possible to separate true change from questions of measurement reliability. Thus, although high test-retest reliabilities are desirable for dealing with problems of measurement error, Heise (1970) has used simulated data and path analysis to suggest that many causal inferences can be made correctly when the indicators are only moderately reliable.

Another problem concerns the concept of "turnover." This concept may lump together such issues as rotation, promotion, recruitment to new positions, release, and retirement with such methodological problems as nonresponse. Careful specification is necessary, since this nonresponse can pose problems for several reasons. If nonresponse varies among different categories of subjects, questions of internal validity and of additional explanations for particular substantive results are raised. Thus, when the nonresponse rate is markedly greater for a control group than for an experimental group, differences between the two groups are accentuated. The possibilities that a change in the experimental group could be due to various factors other than the experimental event become harder to reject the more disparate the control and experimental groups. Campbell (1971) has suggested some methods for estimating the direction and maximum possible magnitude of these selection effects.

From an operational point of view, the problem of nonresponse in short time-series studies can be divided into two components: some individuals will be "lost," while others will refuse to respond. When a second wave of data collection is attempted, there will be some respondents who cannot be located. These individuals may have moved outside the boundaries of the system as defined by the researcher. An inability to locate respondents often suggests too narrow a research focus. Individuals' refusal to cooperate is a possible problem with regard to all the designs considered here. The rates of refusal might differ between different groups (between different experimental groups or between experimental and control groups) depending upon whether or not the members of the group have had a favorable or unfavorable experience.

These designs all depend upon adopting an "actuarial" approach, one of keeping systematic records across time. To help untangle various confounding influences, individual mobility as well as changes in the characteristics of settings, roles, and individuals must be traced. Excellent research opportunities are inherent in the "high rate of transfer" situation. But organizations vary substantially both in the extent to which records are kept at all and in the researcher's degree of access. Organizations may insulate themselves from honest evaluation by not keeping such records or by denying scholarly access to them.

The same kinds of approaches are relevant to the related issues of shifts in measures of central tendency and dispersion and of changes in stability of variables over time. The shift in the measures of central tendency and dispersion poses difficulties for analysis at all levels of measurement—from nominal to interval. Problems exist because such shifts can alter the size of the measure of association in complicated ways. Several approaches to this problem have been suggested.

Mosteller has proposed a different measure of association for the analysis of contingency tables. This measure, the cross-product ratio, is invariant under row and column multiplications.

> Think of a contingency table as having a *basic nucleus* which describes its association and think of all tables formed by multiplying elements in rows and columns by positive numbers as forming an equivalence class—a class of tables with the same degree of association. (1968:4)

Figure 2.1a gives an example of the basic 2×2 table, in which a, b, c, and d are counts of the number of cases in each cell. Figures 2.1b, 2.1c, and 2.1d provide examples with hypothetical data. The Pearson correlation and cross-product ratio are presented underneath each contingency table. The cross-product ratio is generalizable to contingency tables of any size; it is described in detail in Mosteller's article.

If ordinal or interval data are categorized in an appropriate manner, the cross-product ratio can be used as a measure of association. This measure has the disadvantage of being based on a lower level of measurement (nominal). The distortions produced by this technique, as compared with the assumption of higher levels of measurement (interval) for weaker data, are not clear. A rather different approach to this problem of change in category or level has been proposed by Bohrnstedt (1969). His technique is based on the Pearson correlation, specifically designed to take into account, through statistical controls, initial states of relevant variables. In analyzing two or more variables over time, partial correlation coefficients might be used in an effort to remove autocorrelation effects. Thus, a correlation between variable A at time 1 and variable B at time 2 ($r_{A_1B_2}$) would be calculated most appropriately by controlling for the level of variable B at time 1 ($r_{A_1B_2 \cdot B_1}$). Such corrections are important because the relation-

FIGURE 2.1. Examples of 2 × 2 Tables, with Row and Column Multiplications

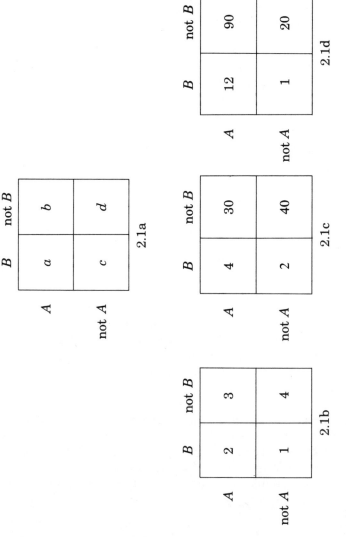

	B	not B
A	a	b
not A	c	d

2.1a

	B	not B
A	2	3
not A	1	4

2.1b

	B	not B
A	4	30
not A	2	40

2.1c

	B	not B
A	12	90
not A	1	20

2.1d

Pearson product-moment correlation (Bivariate normal)	$r = 0.35$	$r = 0.29$	$r = 0.25$
Cross-product ratio	$\alpha = 8/3$	$\alpha = 8/3$	$\alpha = 8/3$

Source: F. Mosteller, "Association and Estimation in Contingency Tables," *Journal of the American Statistical Association,* Volume 63, 1968, p. 3. Reproduced by permission of the author and the publisher.

ships between A_1 and A_2, and between B_1 and B_2, may change differentially over time. As discussed later, zero-order cross-correlations $(r_{A_1B_2})$ are difficult to use for causal inference when the stability of the relevant variables differs substantially.

Certain fundamental concepts and the form of notation used in this paper are summarized here. Variables $(A, B,$ etc.) are measured at times 1 and 2 $(t_1, t_2,$ etc.); subscripts identify the time of measurement with the variable $(A_1, A_2, B_1, B_2,$ etc.). There are individual units $(I-$respondents) who can be located in particular settings (S) and individual units and settings studied at given times which are identified by subscripts $(I_1, I_2, S_1, S_2,$ etc.). At certain times the individual units studied at t_1 (I_1) will be the same as those studied at t_2 (I_2); and in such circumstances, the equivalency will be noted $(I_1 = I_2)$. Events also occur which have implications for the study of both units and variables. Furthermore, they may affect individuals' relationships with their settings and, consequently, research designs. These designs are discussed below.

PANEL DESIGNS

Panel designs — the collection of data on at least two variables from at least two points in time — can reduce the threats to reliability and validity inherent in data collected from one point in time (Campbell and Stanley 1963). The idea here is that a cause will be more closely associated with a subsequent effect than vice versa. The basic cross-lagged panel model is presented in figure 2.2. When lagged cross-sectional correlations are compared in panel studies, a number of hypotheses are in competition. There is the null hypothesis that A and B are unrelated. Rozelle and Campbell (1969) have suggested four additional hypotheses:

  (i) Increases in A increase B, and decreases in A decrease B.
  (ii) Increases in A decrease B, and decreases in A increase B.
  (iii) Increases in B increase A, and decreases in B decrease A.
  (iv) Increases in B decrease A, and decreases in B increase A.

If the cross-correlations $r_{A_1B_2}$ and $r_{B_1A_2}$ differ significantly, a separation can be made between the two pairs of hypotheses. Where $r_{A_1B_2} > r_{B_1A_2}$,

the effect of
(i) increases in A increase B the effect of
         or   } is greater than { (iii) increases in B increase A
(iv) increases in B decrease A or
                                             (ii) increases in A decrease B

When the horizontal correlations (the over-time correlations for variables A and B) differ substantially, Bohrnstedt's (1969) pro-

FIGURE 2.2. Possible Correlations among Variables A and B at Times 1 and 2

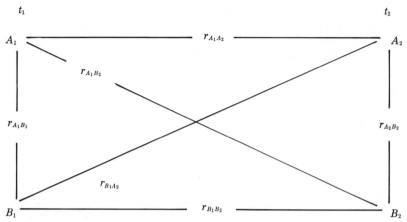

$r_{A_1B_1}$ and $r_{A_2B_2}$ are synchronous correlations.
$r_{A_1A_2}$ and $r_{B_1B_2}$ are autocorrelations.
$r_{A_1B_2}$ and $r_{B_1A_2}$ are cross-lagged correlations.

posal concerning partial correlation coefficients (i.e., comparing $r_{A_1B_2 \cdot B_1}$ and $r_{B_1A_2 \cdot A_1}$) and Mosteller's cross-product ratio could usefully supplement the product-moment correlations. This would provide a control for changes in the central tendency and dispersion of the variables over time.

Other sorts of plausible hypotheses have been suggested. Howard and Krause (1970) believe that "a cause may be either a state (static) or a change of state (dynamic) and the same is true for an effect." More generally, they suggest that causal propositions may take four different forms (static-static, static-dynamic, etc.) even when the direction of causality is specified. A "static-dynamic" interpretation seems appropriate in panel studies. Thus, it might be better if the causal statements were phrased according to level (i.e., "high levels of A increase B") without considering how these "high levels" came about. The focus of attention is upon the level of the causal variable rather than upon changes in that level. If $A \rightarrow B$, the question of what has influenced A remains open.

The opportunities for distinguishing among competing causal hypotheses are not as bleak as implied above. Often some hypotheses can be eliminated on the bases of implausibility and/or systematic evidence. The two possible negative relationships suggested by application of the Rozelle and Campbell scheme (i.e., hypotheses of the "increases in A [high levels of A] decrease B"

type) may be rejected when negative correlations between variables A and B are not found (Crano, Kenny, and Campbell 1972). Rejecting negative relationships allows comparison of the two hypotheses: (i) increases in A (high levels of A) increase B, and (iii) increases in B (high levels of B) increase A. Such comparisons are a first step toward establishing preponderant causation.

Several limitations to the basic cross-lagged model have been noted. If the autocorrelations for variables A and B differ significantly, Pelz and Andrews (1964) and Heise (1970) have argued convincingly that the zero-order cross-correlations will be misleading. As mentioned earlier, the use of partial cross-correlations represents one way to try to take measurement instability into account.

Going beyond the analysis of two variables at two points in time is also desirable. Two-wave, two-variable data are hindered by the fact that assymetrical cross-lagged correlations may be explained as violations of stationarity. Stationarity assumes that the causal parameters of the measured variables do not change and "that the common factor structure of the tests employed at both points in time remains constant" (Crano, Kenny, and Campbell 1972:267). If the underlying factors are the same, the synchronous correlations $(r_{A_1B_1}$ and $r_{A_2B_2})$ should — within the limits of sampling error — be the same.

If these synchronous relationships are not the same, problems of differential reliability, changes in variance, and changes in test specificity must be confronted. Here again, the use of partial correlations (or cross-product ratios) should supplement a reliance upon zero-order product-moment correlations. Additional variables help to specify the kinds of shifts which may have taken place. Kenny (1973) has been analyzing a number of models based on synchronous and lagged common factors. Treatment of these problems goes beyond the scope of this discussion and involves techniques for correcting for changes in the interrelationships among measured variables (Crano, Kenny, and Campbell 1972). After correction procedures, new cross-lagged correlations accounting for shifts in communality can be compared. Asymmetry between corrected cross-lags — both zero-order and partial — may be considered evidence for causation, although the problem of possible shifts in the influence of unmeasured variables remains.

These same problems have been approached also by means of causal path analysis. Heise (1969, 1970) has worked with various combinations of time points and variables, while proposing a causal model suitable for two-wave, two-variable data. He notes that "the time-ordering of the data allows several of the logically possible paths to be discarded" (1970:4); i.e., A_2 is not a determinant of A_1 or B_1. By assuming that causation does not occur instantaneously, Heise can eliminate four other paths, i.e., A_1 does not cause B_1 or vice versa. "Four paths are left as possible connecting links, and the values of the path coefficients for these paths can be estimated from the observed correlations" (Heise 1970:5).

To summarize Heise's observations (1970:5–6), A_1 and B_1 are predetermined (exogenous) variables which affect other (endogenous) variables in the system but are not affected by them. The true correlation between these predetermined variables is represented by the curved line and $\rho_{A_1B_1}$. A_2 and B_2 are endogenous or dependent variables which may be affected by A_1, B_1, or outside disturbances which occur between measurements. The aggregation of these residual variables, these outside disturbances, is represented by u_A and u_B.[2] The path coefficients $(P_{A_2A_1})$ and the correlation between the predetermined variables stand for the basic parameters of the system and are unknowns to be estimated from empirical information.

As noted by Heise, the path model in figure 2.3 can be used

> to write a series of equations expressing the value of each correlation between variables as a function of the basic parameters. These equations then could be solved to express the parameters as functions of the correlations, thereby providing estimating formulas. (1970:7)

As discussed earlier, Land (1969) has provided a general introduction to this technique, while Heise (1970) treats in some detail how to generate estimating formulas and path coefficients.

The path analysis approach outlined by Heise (1970:10–11) makes several sets of assumptions which are shared by the cross-lagged panel approach, including the assumptions of linear regression (linearity, homoscedasticity, noncollinearity) and assump-

---

2. Exogenous variables are predetermined variables whose total variation "is assumed to be caused by variables outside the set under consideration." In endogenous or dependent variables, the total variation "is assumed to be completely determined by some linear combination of the variables (exogenous or other endogenous) in the system." In path analysis, "a residual variable, which is assumed to be uncorrelated with the set of variables immediately determining the variable under consideration and to have a mean value of zero, is introduced to account for the variance of the endogenous variable not explained by measured variables" (Land 1969:6).

FIGURE 2.3.   A Path Analysis Approach to Panel Data

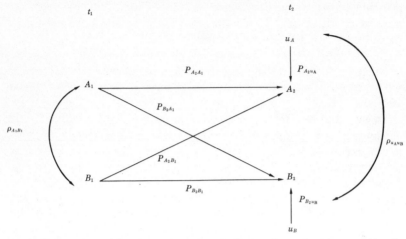

$P_{A_2A_1}$ is the path coefficient to A_2 from A_1. The path analysis convention is to note the influenced variable first in the subscript.

$\rho_{A_1B_1}$ is the true correlation between A_1 and B_1. (In practice, it will be estimated from the empirical correlation $r_{A_1B_1}$.)

u_A and u_B are residual variables, outside disturbances affecting variables A_2 and B_2, respectively.

Source: D. R. Heise, "Causal Inference from Panel Data." In E. F. Borgatta and G. W. Bohrnstedt, Eds., *Sociological Methodology 1970*. (San Francisco: Jossey-Bass, 1970), p. 6. [Notation modified]. Reproduced by permission of the author and publisher.

tions concerning the generality of the causal dynamics. Heise's path analysis model makes additional "assumptions about the timing of causal effects and of measurements, and assumptions concerning extraneous sources of variance" (1970:12). Many of these latter assumptions have been tested by Heise (1970) and Pelz and Lew (1970) using simulated data with known characteristics. Given varying amounts of measurement imprecision and "noise" from outside disturbances, the model performed rather well. This approach is particularly important because it represents an effort to estimate causal parameters, or paths, as well as to establish preponderant causation. The possibilities of reciprocal causation are treated explicitly.

In a related approach using path analysis, Duncan (forthcoming) has explored problems of measurement in panel studies by developing "a general model for the causal linkages that may be present in a set of 2W2V [two-wave, two-variable] data." Particular models are derived from the general model by assuming that certain paths and correlations are equal to zero. Duncan's models explicitly deal with problems of measurement error and questions of simultaneous versus lagged causation, whereas the latter are not treated by Heise. Assumptions about causation can be partially tested by reversing the paths in these models.

The newer approaches to panel data somewhat modify the closed system perspective implicit in the basic cross-lagged correlational approach. In the path models of Heise (1970), influences other than the variables explicitly being studied enter into consideration, but only as residual disturbances. Clearly, these implicit variables are not really considered part of the system. The factor analysis perspective developed by Kenny (1973) emphasizes that observable variables are only imperfect measures of underlying factors. If the relationship between observable variables and underlying factors does not remain constant, corrections in lagged correlations must be made. Thus this formulation is sensitive to changes in relationships between explicit variables and unmeasured factors, and a more open system perspective is achieved.

Although the formulations outlined here will undoubtedly receive increasing attention, advanced data analysis is only a partial solution to the problem of untangling causal relationships. Collecting data using multiple indicators and sampling more time points will help in developing more appropriate causal models. Here research design can aid in specifying causal patterns, but, as Blalock (1970) has stressed, assumptions about measurement error will influence the sophistication of the causal models which can be tested.

Panel Designs — An Example. Data from several points in time could be used to improve the well-known Miller and Stokes (1963) study of representation. Miller and Stokes were concerned with the causal influence of constituency (setting) and congressional opinions on congressional roll-call voting. For the moment, consider Congress as a closed population without retirement and without the defeat of incumbents. The treatment of replacement designs will deal with the implications of these latter factors.

Although published research has concentrated on cross-sectional analysis, the Congress-constituency data include several possibilities for panel analysis. The opinions of a varying number of citizens (the mean seems to have been about 17) sampled within each constituency were averaged to produce the constituency opinion for each of three issue areas — civil rights, foreign policy, and welfare. Figure 2.4 presents data on the welfare issue area.

FIGURE 2.4 Approximate Correlations between Average Constituency Opinions and Congressional Roll-Call Votes—Welfare Issue Area

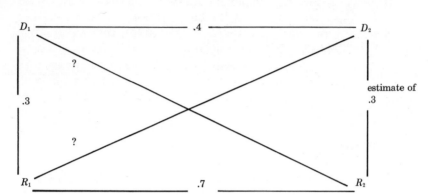

D_1 = average constituency opinion at time 1; D_2 = average constituency opinion at time 2.

R_1 = representative roll-call voting position at time 1; R_2 = representative roll-call position at time 2.

The data are from the 85th and 86th Congresses. Setting and role are held constant; losses due to "experimental mortality" are not considered. Corrections for attenuation have not been made.

Symbols for mean district opinions (D) and congressional roll-call votes (R) are shown, while subscripts denote the time of data collection (Converse 1964; Clausen 1967).

Cross-lagged correlations and path coefficients would help in making causal inferences about the direction of influence between congressman and constituency. Although such data have not yet been published, the magnitudes of the synchronous and over-time correlations make negative cross-correlations unlikely. With the Congress-constituency information, the alternative hypotheses of an increase in D decreasing R and vice versa can be eliminated from consideration. Negative cross-correlations might result from congressional measures which were subsequently unpopular, but the data under consideration did not seem to follow this pattern.

Longitudinal information on the settings at times 1 and 2 have not been made available, but other setting variables might have influenced the congressmen. The possibility of more general causes needs to be treated. Some of the threats to validity due to history, maturation, and testing could be explored, using additional data available from the Survey Research Center's samples. Moreover, further analysis of additional variables — congressional attitudes and congressional perceptions of district opinions — would help in exploring alternative influence patterns and various common factor models.

Finally, both panel analysis and an interrupted time-series perspective are most relevant for exploring the limited data on constituency attitudes and the more accessible longitudinal information on congressional roll-call votes. Since interview data from representatives appear available for only two time points (1958 and 1960), methodologies suitable for handling more limited data from longer time series are most appropriate.

ROTATION DESIGNS

The rotation design shares many similarities with the field experiment. Both designs seek to gather information from ongoing situations where the type of stimuli applied to the subjects are at least partly under the control of others. But in the field experiment, individuals can be assigned to roles and settings according to the researcher's goals; thus, the researcher can influence who gets what stimuli when. This opportunity to manipulate the situation is typically unavailable in ongoing social situations. Researchers must try to introduce something like experimental design into the data collection procedures.

Randomization provides the best research designs, controlling for variables which characterize the respondents (Hilton 1972:47). One set of variables is randomized with respect to another set of variables. Obviously, a set of individual variables remains linked with each individual, and a set of situational variables is linked with each situation. But a rotation design does suggest a strong way to estimate the relative contribution of situational and individual variables.

Given a randomization of individuals among settings, the data collection might be diagramed as follows:

$$O_1 R \; O_2$$

where O = observation and R = randomization. Data can be collected from a particular social situation, although different individuals may be present in the two time periods. Of course, similar data should be collected in each situation and in each time period.

Such a design is especially appropriate for the study of individuals in organizations with high rates of transfer. Military or civilian administrators are likely to be reassigned from one role or setting to another. Nurses and foremen may be rotated from one work situation to another. In most cases, evaluating the performance of such individuals would be most desirable for policy purposes.

Different correlational patterns can be linked with various causal models, but the assumptions differ from those used in panel analysis. In this design, time precedence — the idea of "temporal asymmetry to cause" — is modified. Time-series data are necessary to separate the sets of variables, but within any one time period the analysis is cross-sectional, admitting the possibility of simultaneous causation. As is seen in figure 2.5, if role or setting were an important influence on individual behavior, high correlations between situational variables and individual variables would be expected at both t_1 and t_2, both before and after individuals were rotated among settings. But if setting were less important, relatively low correlations would be expected. Such predictions could be made regardless of whether or not data on the same individuals were collected at t_1 and t_2.

If panel data were available, individual attitudes and behaviors at t_1 could be compared with those at t_2. These horizontal correlations are designated by the dotted lines in figure 2.5. If certain individual attitudes and behaviors are primarily determined by the situation, the correlations between an individual variable's measurement at t_1 and this variable's measurement at t_2 (after rotation) should be low. Moreover, relationships should be higher for individuals rotated between similar situations at t_1 and t_2 than for those rotated between dissimilar situations. On the other hand, when individual variables are the primary determinants of certain attitudes or behaviors, a particular attitude measured at t_1 will be correlated with itself measured at t_2. In this case, the type of rotation will not materially affect the magnitude of the relationships; correlations should be about the same whether the respondents were rotated among similar or dissimilar situations.

The logic of the rotation design should be clear. However, the researcher working from a data bank or old records is likely to have incomplete data due to nonresponse of various types, new entrants in old settings, old respondents in newly created settings, and so on. In such circumstances, the researcher (and his com-

FIGURE 2.5.   Analysis in Rotation Designs

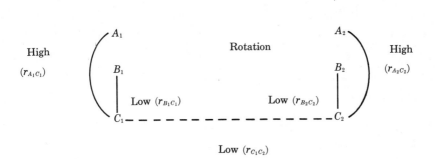

Causal Pattern: Setting Variables (A)
Affect Individual "Dependent" Variable (C)

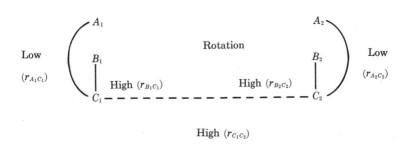

Causal Pattern: Individual "Independent" Variables (B)
Affect Individual "Dependent" Variables (C)

A = setting variables
B = individual variables presumed to be independent
C = individual variables presumed to be dependent

Dotted lines indicate additional correlations relative to the analysis when panel data are available.

High and low correlations are indicated next to the appropriate lines. Thus $r_{A_1C_1}$ = correlation between A_1 and C_1.

Models employing path analysis and taking underlying factors into account are being developed.

puter) may reconstruct the design in terms of two types of matches; he may match the social situation and he may match the individual. For illustrative purposes, the social situation is referred to here as "setting," but it might just as appropriately be called "role."

Type of Match	Situations Studied at t_1 and t_2	Respondents Studied at t_1 and t_2
Setting	Identical	"Random" changes between t_1 and t_2
Individual	"Random" changes between t_1 and t_2	Identical

These matches are presented in figure 2.6. When such data are analyzed in terms of the synchronous relationships between situations and individuals, the results should be independent of the match or matches selected. This provides one check on the internal and external validity of the data.

FIGURE 2.6.   Rotation Design: Setting and Individual Matches

t_1

S_1

t_2

S_2 $(= S_1)$ Setting or role is the same at the two points in time.

I_1

I_2 Different individuals at the two points in time.

Setting Match

t_1

S_1

t_2

S_2 Setting or role is different at the two points in time.

I_1

I_2 $(= I_1)$ Same individuals at the two points in time.

Individual Match

All the settings and all the individuals studied at t_1 may not be available at t_2.

The logic of the rotation design differs from that of the panel design. As diagramed in figure 2.7, the former design shows random reassignment (R) as given, using this fact to investigate some of the relationships among factors A (situational variables), B (individual "independent" variables), and C (individual "dependent" variables). This design implies that A, the situational

FIGURE 2.7.   Rotation and Panel Designs

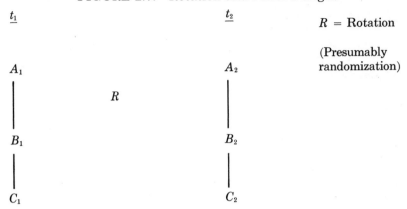

Variables in Rotation Design

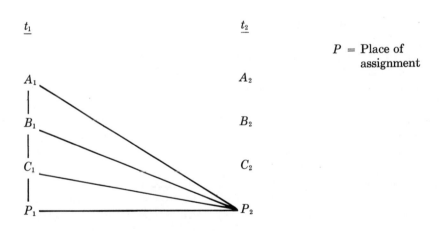

Variables in Panel Design

A = setting variables
B = individual variables presumed to be independent
C = individual variables presumed to be dependent

variables, are considered to be isolated from influence from B and C; the emphasis is upon comparing the effects of A and B on C.

A panel approach would consider the place or position of assignment (P) as a possibly dependent variable. Thus P—which could be operationalized in terms of formal rank or desirability of a particular setting – is shown as a dependent variable in figure 2.7. Even if, for purposes of simplification, A – the situational variables – are taken as isolated from influence by B and C, this conceptual scheme suggests the influence of A_1, B_1, and C_1 on P_2, the place of assignment at t_2. Reassignment becomes an event to be explained rather than an event taken as given. The possible inter-relationships among variables are multiplied, and, if assignment is not random, controls for unmeasured outside factors are not present.

In a quasi-experimental design it is impossible to be sure that reassignment and rotation are random. The situation becomes one of trying to explain variance in the reassignment process. The variables are analyzed in an effort to understand how individuals and settings are sorted out. Such explanatory efforts should use all available data, but, because completely unknown factors may affect individual assignments, they can only be partial.

Testing for randomness of rotation involves standard bivariate and multivariate techniques. A number of social background characteristics of individuals (the B variables in the diagram) are essentially constant and might be predictors of assignments; running lagged and cross-sectional correlations, respectively, would be appropriate. Individual attitudes or orientations (C variables) might affect position or place of assignments. Finally, an individual's position at t_1 (P_1) could influence his position at t_2 (P_2).

Data from the same individuals at several points in time aid in detecting various deviations from randomness. As with the straight panel study, the "loss" of particular categories of individuals may provide information about career patterns and suggest changes in research design. Departures from randomness in reassignment complicate separating the effects of setting variables from those of individual variables. In this situation, correlated independent variables pose a constant danger to interpretation. Testing for a single-factor model may help prevent a misspecification of causal relationships (Brewer, Campbell, and Crano 1970).

The rotation and panel designs as well as the other designs treated here differ significantly from true experimental designs. Rather than assigning subjects to various treatment and control groups, the quasi-experimental researcher is striving to impose order on the relatively chaotic "real world." Dividing the sample according to one or more criteria and assuming random rotation for only one subset of the data might be desirable. Both panel and rotation analysis can be applied to the same data set. Figure 2.8

FIGURE 2.8.   Idealized Rotation and Panel Designs

Rotation Design	Panel Design
Place of assignment appears to be random; available variables cannot predict assignment.	Place of assignment determined by one or more variables internal to the research, or by more general cause.
$P = f(u)$	$P = f(A_1, \ldots A_n, B_1, \ldots B_n, \text{etc.})$

P = place of assignment
A, B = different types of variables
u = random variable

shows how these two analytical strategies might be considered as opposite ends of a continuum.

It is difficult to specify in operational terms the point (in terms of statistically significant relationships or variance explained) at which the benefits of randomization must be presumed lost. Since many of the checks for randomness involve lagged relationships, there is no contradiction between strategies of analysis. For intermediate cases characterized by both a substantial random component and some meaningful nonrandom factor(s), data might be presented according to both models. Such an approach would obviate having to make calculations – perhaps based on a concern for Type 1 and Type 2 errors – as to which model is appropriate.

More complicated formulations based on the rotation design are possible. A linear regression approach would seem appropriate for some efforts at predicting place of assignment, but the assumptions of the linear model may not always be met. Reference to the econometrics literature (e.g., Johnston 1963) would be helpful here. Another line of thinking could be to develop the similarities between the rotation design and the Latin square design. Each role or setting might be considered a treatment; cumulative effects of such "histories" on individual attitudes or behaviors are possible. Because relatively little attention has been paid to the rotation design, there are a number of possibilities for development.

Finally, the implications of these designs for evaluation research should be reemphasized. Although occasional field experiments on performance in different settings have been undertaken (Feldman 1937), little research utilizing this methodology has been done. More generally, studies based on strong designs – designs which eliminate a number of plausible hypotheses—have seldom been used to study the performance of police, hospital personnel, foremen, and so forth.

Rotation Designs—An Example.   An extensive discussion of a complicated rotation design with missing data is provided in the appendix. Lieberman's (1956) study of the effects of work role on individual attitudes will be treated here as a study with some characteristics of the rotation design. As summarized by Katz and Kahn:

> [Lieberman] was able to measure the perceptions and attitudes of employees in two appliance plants three times during a period of three years: once when all were rank-and-file workers, a year later when 23 had become foremen and 35 had been elected stewards, and two years later still when about half of the new foremen and stewards had reverted to nonsupervisory jobs and half had continued in their new roles. . . . On becoming foremen, Lieberman's subjects tended to report more favorably about the company as a place to work, to be more favorable in their perceptions of top management, and to endorse the principle of incentive pay. Those men who became union stewards became, according to their responses, more favorable toward unions in general, toward the top officers of their own union, and toward the principle of seniority rather than ability as a basis for wage payments. Those foremen and stewards who subsequently returned to the worker role tended also to revert to the perceptions and attitudes of workers; those who remained as foremen and stewards showed more sharply as time passed the kinds of differences described above. (1966:189)

These data from three time points "argue strongly for a causal relationship between the office an individual occupies in an organization [foreman, steward] and his expressed attitudes on job-relevant matters" (Katz and Kahn 1966:189).

How can the logic of the rotation design be applied here? Individuals who were demoted (at t_3) might be considered as having been rotated back to another setting. The change in attitudes among these individuals helps deal with the hypothesis that a common factor might have been underlying both the promotion and the attitude change processes.

On the other hand, promotion (and probably demotion) were not random; the selection was biased. Both future foremen and future stewards "were more ambitious, more critical and less unquestioningly loyal to the company than were the workers destined to become neither foremen nor stewards" (Katz and Kahn 1966:189). For these types of data, an extensive panel analysis would prove useful.

The influence of completely unknown variables obviously cannot be ascertained. Because the situation is not truly experimental—because individuals have not been randomly promoted—the effects of role alone are not precisely known. The particular characteristics of the promoted workers probably interacted with work role as determinants of attitudes. In future analyses of this type, the importance of these background variables needs to be explained as fully as possible.

More generally, when enough respondents are available, the researcher might control for differences in characteristics of individuals moving into various situations. But, as has been shown by other investigators, controls provide weak substitutes for true experimentation. When groups of individuals have different means, even basic controls—and especially attempts at matching—are hindered by regression artifacts.

Finally, as is discussed in the following section, this study could also be conceptualized in terms of a pretest-posttest design. It is not clear whether the act of promotion (an event) or the new work role was responsible for the attitudinal changes. Indeed, such different causal variables may be impossible to separate without either (1) individuals who experience one (promotion or higher status role) without the other, or (2) strong assumptions about the length of time that attitude change due to change in state (i.e., due to promotion) persists.

PRETEST-POSTTEST DESIGNS

Thus far this chapter has dealt with the relationship between events and short time-series data in a restricted fashion. When noted at all, events have been considered only with regard to rotation or as threats to validity (i.e., in terms of history). Although political and social changes are likely to affect both rotation and turnover, they are important for other reasons as well. Experiencing a given event and/or participation in a particular program may well change individual attitudes and behaviors. The group subject to this event can be thought of as the treatment group. In order to deal with the main effects of such threats to validity as history, maturation, testing, and instrumentation, a control group is necessary (Campbell and Stanley 1963). In a true experiment, variable *A* in the pretest-posttest framework is completely exogenous. The event or treatment is outside the control of the individuals who will experience it, affecting some individuals but not others in a random fashion. In a quasi-experiment, random assignment to treatment and control groups from a pool of individuals is not possible. But, although selection may interact with other factors, a nonequivalent control group can help deal with a number of threats to interpretation of the data. Pretest-posttest designs using more than one control group are discussed in detail elsewhere; problems of differential growth rates (selection-maturation interaction) remain among the most critical threats to the validity of this design (Campbell and Erlebacher 1970).

Following Campbell and Stanley (1963), this design can be presented as:

$$t_1 \qquad\qquad\quad t_2$$

$$O_1 \qquad X \qquad O_2 \quad \text{(treatment group); } X = \text{``experimental''}$$
event
$$O_1 \qquad\qquad\quad O_2 \quad \text{(nonequivalent control group)}$$

Problems of self-selection are central to both the panel and pretest-posttest designs. For the panel design, this problem is "internalized"; the presence of any underlying common factor must be considered as a rival explanation for a hypothesis of the type, "a change in A causes a change in B." In the pretest-posttest designs, one or more control groups which have not experienced event A should be used. In a quasi-experimental situation, such control groups will not be precisely equivalent to the "experimental" group, but efforts can be made to make the groups as equivalent as possible (Campbell and Stanley 1963).

Pretest-Posttest Designs — Two Examples. A certain rephrasing of the problem illustrates the continuity between designs emphasizing "events" or "interventions" and panel designs. In my Turkish data, many respondents working in the Ministry of Interior (at t_1) moved to new ministries which paid more and promised better career opportunities (at t_2). Such movement can be conceptualized in several ways. If some individuals working in new ministries were available at t_1, a panel design would seem appropriate.

However, as suggested by the generalized diagram in figure 2.9, a variant of the pretest-posttest design might be adapted to the data. In the Turkish case, the variable A might be conceived of as change in place of work or as change in work experience which might affect a dependent variable B, such as job satisfaction.

FIGURE 2.9.   Pretest-Posttest and Panel Designs

t_1 t_2

A = presumed independent variable

A

↓ B = presumed dependent variable

B_1 B_2

Control groups may or may not be present,
altering the strength of the design

Variables in Pretest-Posttest Designs

t_1 t_2

A_1 A_2
                A, B = variables
B_1 B_2

Variables in Panel Design

With the Turkish data, administrators selected themselves for the "event" (moving from one ministry to another). Because such selectivity was directly related to experiencing event A, the pretest-posttest aspect of the design is weakened. Such self-selection suggests the need to determine the predictors of this mobility; this has been done elsewhere (Roos and Roos 1971). The Turkish data were also analyzed extensively using the panel paradigm.

This example illustrates in a rather striking way the common social science problem of defining the relevant variables. With the pretest-posttest design, it is especially important to distinguish between an event which merely shuffles individuals from one setting to another and an event which may be a cause in itself. Just as with the rotation design, the movement of respondents from setting to setting can often be presumed to be noncausal. In many pretest-posttest designs, such shuffling divides the sample into before-after segments and makes it possible to investigate change in the dependent variable. In order to judge what event or other variable is to be taken as a direct cause, a number of variables which may have affected one another over a given period of time should be considered. This may lead the researcher back toward panel analysis.

Such a discussion is directly relevant for policy research. Both Shingles' research (chapter 7) and Armor's (1972) study on busing used nonequivalent control group designs. Armor's design might be briefly discussed here, since it represents one of the best evaluations of an ameliorative social program. In the Armor study the causal role of the bus ride per se is presumed to be minimal; what seems important is the difference between the settings at t_1 (mostly black schools) and those at t_2 and t_3 (mostly white suburban schools). Some commentators note the length of time lost by relatively long bus rides; such statements suggest that riding the bus could in itself have a (negative) impact on grades. On the other hand, the shared experience of riding the bus might increase group solidarity and change racial attitudes.

Taking busing as the experimental event, Armor (1972:97) noted that an "adequate" control group is one of two types: (1) a group of nonbused black students who are reasonably comparable to the bused black students, and (2) a group of white students in the same school as the bused black students. In the latter case, the effects of integration are revealed in the changes in the black/white differential for the measure in question. In an ingenious fashion, siblings of the bused students were tested in an effort to maximize the equivalence of the (first) control group and the treatment group. This controversial study called into serious question the academic effects of busing, while showing some unintended effects such as increased hostility between races. Without going into greater detail here, the design aspects of this research should be commended. Despite some difficulties with nonresponse and selection bias in the loss of respondents from the (first) control group, the study provides an unusual example of methodologically sophisticated policy research.

The pretest-posttest design may suffer from its emphasis on the causal effect of the event (variable A). As already noted, a microanalysis of this event and other events accompanying it is clearly needed. If sufficient numbers of black students were, by reason of residence, attending schools to which other black students were bused, a more extensive analysis of the impact of busing to an unfamiliar social setting might be possible.

A related strategy would be one focusing on setting variables at t_1 and t_2. In what type of setting do the most significant changes take place? Can classroom, teacher, and school variables be measured in continuous fashion so as to permit panel analysis of these data? More generally, the question is whether exposure to an intervening event measured in discontinuous fashion is important or whether changed levels of other variables are exerting a causal influence.

These research suggestions point out that it often will be possible to try several approaches with a single substantive problem or even with a single data set. In quasi-experimental situations, the relationships among most variables will not be known until the data analysis stage. The directives for the optimal combination of design and analysis might be as follows: make the design or designs as strong as possible, keep the assumptions clearly in mind, and then use analytical techniques to try to isolate remaining sources of variation. If multiple approaches to the same data set are possible, statistical analysis may help the researcher to decide which approach is the more appropriate.

Along these lines, some of the similarities between the rotation design and pretest-posttest designs may be usefully noted. As quasi-experiments, both represent weaker variations on true experiments, lacking truly random assignment to experimental and control groups. Both of these quasi-experiments are characterized by similar threats to internal validity. In these designs, exposure to a particular experience may be internally, rather than externally, determined. The designs differ with regard to the role of the intervening event, although the pretest-posttest design requires a better description of assumptions in this regard. Equally important, in the rotation design some individuals move into positions vacated by other individuals. They move both ways, which is not the case with the pretest-posttest design. With both designs, many of the checks on selective assignment are statistical and necessarily incomplete. The more data available, the more extensive these checks can be.

At this point the discussion might be connected with the previously mentioned stress on collecting data from numerous points in time. Such longer time-series data aid in handling questions of measurement error and generalizability. Because "all relationships tend to weaken with time" (Campbell 1971:106), longitudinal data are also valuable in conjunction with pretest-posttest designs. Temporal erosion of the correlation between one variable (such as exposure to a given event or treatment) and another (such as an attitudinal or behavioral measure) can complicate interpretation of treatment effects when there is a pretest correlation between these variables, i.e., a selection effect. Campbell (1971) has discussed this problem in some detail.

REPLACEMENT DESIGNS

An "open system" perspective is essential for taking into account movement into and out of social roles. There is no single replacement design in the sense that typical designs have been described above. Replacement designs constitute an attempt to apply the logic of "open systems" to questions of design. In democracies, leadership succession is more than just "experimental mortality," more than yet another threat to validity. Questions of migration and mobility, of recruitment to and release from important positions, are central for many behavioral scientists.

The issue of selection — of replacement and recruitment — is central to many policy problems. As Campbell (chapter 6 of this volume) has stressed, selection and treatment are often confused. The policy of "skimming" — of helping the most advantaged persons in any pool of individuals from which a social program is supposed to draw — is widespread. Recruiting such individuals may tend to make the program's output "look good," particularly if data are collected from one point in time — after the training or assistance has been finished.

If the researcher believes turnover and migration to be significant, a replacement design will be called for. This design is particularly appropriate when the rank or formal position of the respondents is likely to change rapidly. Replacement designs often emphasize either individuals entering or those leaving the unit being studied. Legislatures where representatives are either re-elected or defeated in their bids for office provide a good example of such situations; electoral victory or defeat determines membership.

This expansion of the framework suggests ideas for utilizing one-time respondents who have left the research setting and are difficult to locate. For example, individuals who have left a busing situation to return to predominantly black schools might be studied in order to assess the long-term effects of the busing experience. Resources invested in tracking down such individuals would probably pay both theoretical and methodological dividends.

Replacement designs can also help investigate plausible rival hypotheses threatening the interpretability of an empirical relationship. An appropriate focus on replacement is likely to lead to the use of several additional groups to eliminate rival hypotheses and strengthen the research design. Panel studies are particularly complicated by the rival hypotheses of changes due to history or maturation.

These threats to validity refer to processes of vital importance. Research involving the collection of data from time points which are even moderately spaced must take into account external events and changes in the level of one or more of the measured variables. More formally, unknown or unmeasured variables are likely to affect the measured variables.

An awareness of such variables may help guide data collection, but additional cautions are necessary. If the processes of measurement can be assumed to have had no effect on the responses, the task is simplified. The collection of new data on new individuals at t_2 would help disclose any effects of history and maturation on presumed dependent variables. Data collected at t_2 from the individuals at the same age or life-cycle stage as those studied in the first wave at t_1 are especially valuable in dealing with history and maturation. Various checks on differential selection of this new group should be incorporated. These younger individuals are presumably occupying roles similar to those formerly held by the panel respondents. Such a design might be diagramed as follows:

t_1 t_2

O_1 O_2 (Panel data — successive observations on the same individuals)

          $O_{2'}$ (Data from individuals the same age as panel respondents at t_1)

This design would utilize three types of information: (1) panel data — information from the same individuals at different stages in their life cycle; (2) age-level data — information taken at different times from different people at the same stage in their life cycle; and (3) cross-sectional data — information on individuals at different stages in their life cycles, but experiencing similar societal events. If historical changes were of major importance, one would anticipate little variation in cross-sectional comparisons but substantial differences for the panel and age-level comparisons. If maturation were important, differences for panel and cross-sectional, but not for age-level, comparisons would be expected. Although the effects of interaction complicate analysis, this design can aid substantially in eliminating plausible rival hypotheses (Roos and Roos 1971).

Replacement processes are equally applicable when individuals move from one role or setting to another. Although tenure provisions may protect administrators from outright release, ineffective or rebellious bureaucrats are frequently transferred from desirable positions while especially competent administrators are promoted. Such actions create vacant slots which are eventually filled by new entrants into the organization. The presence of such vacancies may be a major factor influencing individual career mobility (White 1970).

Variables other than turnover also have implications for design. Situations of organizational expansion or contraction call for some variant of the replacement design. Because the size of businesses and bureaucracies tends to vary, the study of recruitment and release is particularly important in these organizations. A panel design using an individual's rank or position as one variable is useful for such a study, but the problem of new entrants is not solved by this sort of approach. In particular, new entrants may differ from their older counterparts because of either changed selection procedures or self-selection.

Both differential selection and differential mortality pose threats to generalizability in researching situations with high levels of turnover or growth. Once the limits of the unit to be investigated have been defined, the following criteria may help to evaluate the necessity for a replacement design: (1) rate of entry of individuals; (2) rate of exit of individuals; and (3) rate of internal change within the system. Rates of internal change are obviously difficult to measure. Indicators such as the number of new positions in proportion to the total number of positions may provide a partial measure of such changes. Absolute percentage criteria also could be used, but more sophisticated statistical formulations might be appropriate. It should be possible to develop confidence intervals for population values, given different kinds of data about exit, entry, and internal change.

Thus, replacement designs help deal with the substantive and methodological problems of history, maturation, and generalizability. It is not surprising that students of comparative politics (Butler and Stokes 1969; Inglehart 1971) have been particularly concerned with replacement and socialization processes taking place within the electorate. The analysis of peer groups (Evan 1959) is related to this emphasis on replacement, since the study of long-term shifts in attitudes must be juxtaposed with shifts in demographic structure. Here a focus on micro-level data promises major rewards in increased understanding of the larger unit. The following example of a replacement design is also concerned with a political unit: the United States Congress.

Replacement Designs—An Example.   A replacement design would be useful for the study of representative-constituency relationships in the United States House of Representatives. Since the

House remains constant in size, growth factors need not be taken into consideration. In the House of Representatives, two-year terms and periodic redistricting encourage a fair amount of turnover. Although 89 per cent of the members of the 84th Congress were reelected to the 85th Congress in 1956, only 81 per cent of the representatives in the 85th were reelected in 1958.

An "ideal" replacement design is hard to specify, since the design will depend upon the unit being studied. Taking the congressional example and assuming no mortality among congressmen, a replacement design might include: (1) former congressmen who were not returned to office in the election between t_1 and t_2; (2) congressmen who were reelected; (3) new congressmen chosen in the election between t_1 and t_2; and (4) challengers who were not elected between t_1 and t_2. These possibilities are diagramed in figure 2.10.

FIGURE 2.10.   A Replacement Design for the Study of
Representative-Constituency Interaction

t_1		t_2	
	I_2	/Congressmen not returned to office in election between t_1 and t_2 \	Individuals the same as in t_1
I_1 (Congressmen)	I_2'	/Congressmen reelected at t_2 \	
	I_2''	/New congressmen chosen in the election between t_1 and t_2 \	Positions the same as in t_1
	I_2'''	/Challengers not returned to office in election between t_1 and t_2 \	Unsuccessful new competitors for positions

The settings (congressional districts) studied at t_1 and t_2 are taken as the same.

The groups selected for study at t_2 may vary according to the researcher's emphasis. For the study of individual attitude change, it is desirable to resurvey both individuals remaining in the organization and those who have left — in this case, the reelected and the defeated congressmen. If the congressional role is of prime interest, the sample at t_2 might incorporate both the reelected and the newly elected congressmen. Including all leadership groups in the study helps to specify which attitudes tend to be necessary for selection by followers (necessary for being elected to Congress), which tend to be obtained through socialization to the (congressional) role, and which are changed over time, as a result of either individual attitude change or recruitment of new individuals to leadership roles. Indeed, such threats to validity as history and maturation might be handled better by more extensive data collection at t_1. Challengers who were defeated in the election before t_1 would be a desirable group to study, as such data collection would help deal with the generalizability of particular electoral patterns.

SHORT AND LONG TIME SERIES

These short time-series approaches can be compared with those using longer time series. Time-precedence is important for various sorts of quasi-experimental designs. In the interrupted time-series design, data on one or more dependent variables are examined before and after a possibly important event to see if the event affected the measured variable(s). The particular event (variable A — conceived of as the independent variable) precedes the change in level or slope of the dependent variable (variable B). In order to deal with the possibility that both variables have been caused by something else — perhaps a shift in another variable — it is necessary to investigate what has preceded the "important" event.

This issue of the more general cause poses a major threat to the validity of the interrupted time-series design for several reasons. First of all explaining particular events complicates matters significantly; any given event is likely to occur in conjunction with many other possibly important events. Secondly, the possibility of lagged causation — of variable A and/or of variable B — opens up a Pandora's box for consideration of earlier events as potential general causes.

Some of these problems are clarified by comparing the interrupted time-series design and the panel design. Although these designs are discussed by Duvall and Welfling (chapter 3 of this volume), my treatment of them is somewhat different. In order to use an interrupted time-series approach, we need a number of data points — but only one case. Information on one or more dependent variables collected over a period of time from a single case (for example, a nation, state, or individual) can be analyzed.

But the interrupted time-series design is stronger if a number of independent cases — multiple replications — are used. Although

FIGURE 2.11. Interrupted Time-Series and Panel Designs

$$A \ (C \ D)$$
$$\downarrow \ \ \downarrow \ \downarrow$$

Case 1 $B_1 \ldots B_f \qquad B_{f+1} \ldots B_j$ 　　　A = presumed independent

$$A$$
$$\downarrow$$

Case 2 $B_1 \ldots B_m \qquad B_{m+1} \ldots B_q$

$$A \ (C)$$
$$\downarrow \ \downarrow$$

Case 3 $B_1 \ldots B_j \qquad B_{j+1} \ldots B_n$

$$A \ (D)$$
$$\downarrow \ \downarrow$$

Case m $B_1 \ldots B_g \qquad B_{g+1} \ldots B_p$

A = presumed independent
　　variable
B = presumed dependent
　　variable
C, D = other independent
　　variables possibly
　　affecting B

m cases from m units

Interrupted Time-Series with Multiple Replications

Case 1　Case 2　Case 3　　　　Case m

$A_1 \qquad A_2 \qquad A_3 \qquad A_4 \ldots A_m \qquad A_{m+1}$　A, B = variables
$B_1 \qquad B_2 \qquad B_3 \qquad B_4 \ldots B_m \qquad B_{m+1}$　m cases from 1 unit

Panel Design from Long Time Series

Case 1　　Case 2　　Case 3　　　Case m

$A_1 \quad A_2 \quad A_1 \quad A_2 \quad A_1 \quad A_2 \ldots A_1 \quad A_2$　　A, B = variables
$B_1 \quad B_2 \quad B_1 \quad B_2 \quad B_1 \quad B_2 \ldots B_1 \quad B_2$　　m cases from m units

Panel Design from Two Time Points

the important and presumably causal event (variable A) will have occurred in all cases, the circumstances surrounding this event are likely to differ in each case. As is seen in figure 2.11, such variation in surrounding circumstances should permit the ruling out of a number of other events as plausible causes. Moreover, explicit time-series comparisons among cases alert the investigator to deviant or interesting cases (see the use of this technique in Duvall and Welfling [chapter 3]).

As noted earlier, the panel approach uses a number of cases for at least two variables measured at more than one point in time. This may be done two ways. If a sufficiently long time series for two variables were available from one unit, each set of observations could be taken as a case. Thus, with A_t and B_t representing measurements of the relevant variables at the subscripted times, $\{A_t, B_t, A_{t+1}, B_{t+1}\}$ would be case 1, $\{A_{t+1}, B_{t+1}, A_{t+2}, B_{t+2}\}$ would be case 2, $\{A_{t+2}, B_{t+2}, A_{t+3}, B_{t+3}\}$ would be case 3, and so forth. The number of cases would be one less than the number of data points. If, for each case, the temporally earlier measures (the first pair in each set of measurements) are subscripted t' and the later measures (the second pair) are subscripted $t' + 1$, panel analysis can proceed in the methods outlined earlier. Since the cases are not independent, some problems in interpreting various correlation and regression coefficients in terms of statistical significance might be expected.

In general, short time-series data from a number of units are analyzed according to the standard panel paradigms. In such circumstances the number of cases would be equal to the number of units. Here, $\{A_t, B_t, A_{t+1}, B_{t+1}\}$ from unit 1 would be case 1, $\{A_t, B_t, A_{t+1}, B_{t+1}\}$ from unit 2 would be case 2, etc. Data from three or four time points could be analyzed in similar fashion.

From a methodological standpoint, panel designs have been more extensively developed than has the interrupted time-series design. For example, problems such as that of measurement error have been discussed with regard to panels, but have not been treated for interrupted time series. The work of Bohrnstedt and Carter (1971) suggests that measurement error, in that it presents a problem for regression analysis, is likely to pose problems for interrupted time series. Questions of sampling error remain largely unexplored.

The panel approach treats both "independent" and "dependent" variables rather differently than does the interrupted time-series approach. In panel analysis the possibility that the presumably "dependent" variable might influence the presumably "independent" variable is not excluded; the researcher's initial ideas about causality might have been wrong. By way of contrast, in interrupted time-series analysis the independent variable is exogenous to the system being studied, and the dependent variable is not considered as a possible cause of the independent variable.

A final difference between designs concerns the nature of the independent variable. In the interrupted time-series design, the independent variable is a significant event which either has or has not taken place. Only abrupt changes in the dependent variable are noticeable. This "discontinuous" way of looking at the variables can be contrasted with treatment of both independent and dependent variables in continuous terms, as in panel analysis.

Gradual changes in variables can be studied by means of this latter technique. Moreover, because measures of association are available for correlational information, a sense of the magnitude of the relationships—both static and dynamic—is possible for panel data, but lacking for interrupted time-series data. These differences in perspective suggest the complementary nature of interrupted time-series and panel designs. The ability to shift back and forth between different sorts of logic is desirable for the researcher.

Both legislative reapportionment and coalition research can be discussed in these terms. Legislative reapportionment provides only one example from the state politics literature; the other analytical possibilities are numerous. Existing research on state government has characteristically used rather weak designs, concentrating on additive and nonadditive models to predict such government outputs as state expenditures (Jacob and Lipsky 1968). More sophisticated formulations have subtracted (or divided) the dependent variable at t_2 from the level at t_1, then correlated this change with socioeconomic and political variables from one point in time (Strouse and Williams 1972). One important limitation of this technique is, of course, that change in the hypothesized independent variables is not taken into account. The approach is basically cross-sectional, and the possibility that the "dependent" variables might affect the "independent" variable is generally not acknowledged (Bohrnstedt 1969).

Legislative reapportionment has taken place in a number of states, presumably as a consequence of judicial decisions. Although cross-sectional studies of malapportionment have found "few measurable policy differences between states which were well apportioned and states which were malapportioned" (Dye 1972: 246), changes in several dependent variables might be expected from reapportionment. For example, reapportionment might result in changes in both the level and distribution of various government expenditures and services. If data for sufficient time points were available, an interrupted time-series approach would be appropriate. The data points would be split into those before and those after reapportionment. Information from as many states as desired could be analyzed from this perspective.

There are several reasons for approaching these data from a panel analysis perspective. Rather than assuming that a before-after study is the most appropriate, the fairness of legislative apportionment might be considered a more theoretically useful independent variable. Such a decision does modify the substantive focus of the research. The fairness of legislative apportionment—

perhaps operationalized by a measure running from 0 to 1 — is considered as a possible cause of variation in the dependent variable. Differences in degree are significant in the panel design; the focus moves away from the idea of a discontinuous shift in an independent variable. A correlational or path analysis approach, rather than an emphasis on before-after change in the dependent variable, is chosen. Formal reapportionment might be a prior cause of change in legislative apportionment, but this is outside the focus of the panel design.

As noted earlier, the data requirements of the two designs differ; inadequate longitudinal data on the dependent variable(s) might not permit application of the interrupted time-series design. With fewer than five time points, the estimates of the slope of the "before" line are unstable. If a number of cases (states, in the apportionment example) are available, the panel approach, using as many time points as possible, may well be more suitable.

Recent research on political coalitions provides additional examples of the usefulness of a quasi-experimental approach. The findings of an important study by Dodd (1972) illustrate the difficulties of interpretation which typify much social science research. Hypothesized independent variables, party system fractionalization and party system instability, are highly intercorrelated; multiple regression of these independent variables against the dependent variables of cabinet coalitional status and cabinet durability produces little increase over the variance explained by simple bivariate regression analysis. Given such multicollinearity among supposedly independent variables, it is particularly difficult to separate the influence of one variable from another. And, of course, even if the independent variables were not intercorrelated, problems of inferring causation from cross-sectional data would remain.

The data used in Dodd's coalition research were generated from 99 parliaments and 179 cabinets in 16 Western parliamentary democracies between 1918 and 1970. In similar fashion, Browne and Franklin (1973) have explored coalition size and payoffs (in terms of cabinet seats) with data from 114 cabinets in 13 European democracies between 1945 and 1969. Thus, from each country a number of coalitions were studied over a substantial period of time.

The standard cross-sectional perspective leads researchers to pool all coalitions from the relevant sample of countries. As McNemar (1969:254) has noted, this leads to some difficulties in interpreting tests of significance. An alternative approach might be to examine the data country by country. Since the mean number of ruling coalitions per country is 11.3 in Dodd's (1972) study, interrupted time-series and panel approaches might be possible. If the interrupted time-series approach were chosen, the possibly important event (variable A) could be a marked shift in party system fractionalization or party system instability. Circumstances where one variable rather than the other shifted would be valuable in disentangling the effects of fractionalization from those of instability.

A panel analysis approach might analyze these several variables (party system fractionalization, party system instability, cabinet coalitional status, and cabinet durability) country by country through time. At the minimum, this would involve comparing the bivariate cross-lagged correlations $r_{A_{t'}B_{t'+1}}$ and $r_{A_{t'+1}B_{t'}}$, using the panel design for long time series. Such a design would pick up gradual changes, while country-by-country analysis would provide multiple replications. A more standard panel analysis, using each country as a separate case, should also be tried. Various rationales could be used for the choice of t_1 and t_2 points; these points might be selected so as to provide considerable variance in either variable A or B. Thus, the time points might be chosen so as to provide a measure of party system fractionalization (A_1) before a significant shift, while A_2 would be this measure of fractionalization after a shift. Such a selection of time points clearly raises the possibility that other exogenous variables could be responsible for the change in variables A and/or B. This sort of selectivity is best used in combination with other analytical strategies less susceptible to this exogenous factor(s) explanation.

MULTIPLE REPLICATIONS THROUGH TIME

The research designs forwarded here have been taken from several disciplines and substantive areas. All have emphasized change in one or more variables. A switch in focus may suggest still other approaches. A comparative perspective has been introduced through multiple replications; the use of control groups, independent samples, and new cases all relate to this comparative logic. But the analytical focus could be reversed. A more static approach, one emphasizing variation within the sample at a single point in time, might be tried. Despite the numerous threats to internal validity which characterize cross-sectional designs (Campbell and Stanley 1963), multiple replications through time may help deal with such threats.

Simon-Blalock causal modeling is often applied to cross-sectional information in an effort to untangle the myriad relationships among variables *A, B, . . . N* (Blalock 1964). Essentially, this method uses correlation coefficients and regression coefficients, some of which are controlled for when other variables are predicted to equal zero. The path analysis approach builds upon this methodology and has been extensively developed for the estimation of causal parameters.

This approach to isolating preponderant patterns of causation has several valuable aspects:

1. The logic of the method in terms of eliminating implausible causal hypotheses is well defined.
2. This technique is concerned with the amount of variance in the "dependent" variable. Quasi-experimental research focuses upon "statistically significant" relationships without looking at the total variance explained. Better techniques for quasi-experimental researchers are clearly needed.
3. The possibility — indeed, probability — of multiple influences upon one or more variables is stressed.
4. When applied to several data sets, a valuable comparative perspective should result.

Some of the limitations of causal modeling have been noted by various authors (Blalock 1970; Hilton 1972):

1. When used with cross-sectional data, change in the variables, either dependent or independent, is not part of the design. In this sense, the design is not quasi-experimental.
2. Insofar as the cross-sectional data collection involves retrospective questions, measurement error may be very important.
3. Strong assumptions as to the general direction of causality are made (particularly with cross-sectional data) and not tested adequately.
4. Several equally plausible models are likely to emerge from the analysis.
5. A single shared factor and/or measurement error may be responsible for observed correlations, regression coefficients, and path coefficients (Brewer, Crano, and Campbell 1970).

In this connection, Pugh and Hickson (1972:274) have made the important distinction "between data that are cross-sectional in concept and those that are merely cross-sectional in data collection." If the theoretical model used is longitudinal, the temporal order of the variables is likely to be relatively clear — i.e., father's occupation precedes respondent's present occupation. Such temporal ordering serves to reduce the number of plausible models and reinforce statements about direction of causality. Thus, although the Miller and Stokes (1963) congressional data were originally approached by means of causal modeling, there are more threats to the inferences made from that cross-sectional analysis than to those made from the models presented below.

A brief example from my earlier study of Turkish administrators is useful here. In this study panel, data were collected from 241 Turkish administrators in 1956 and 1965. These administrators were graduates of one of the most elite university faculties. Comparative data were collected in 1965 from 69 younger bureaucrats. As is shown in table 2.2, a number of predictors of major course of study and organization of work were strongest for the 1946–1955 university graduates and weakest for 1958–1961 graduates in 1965.[3]

Various causal models were examined using these data and causal inference techniques. Summarizing these results, the formulation presented in figure 2.12 seems appropriate for each of the three data sets. Although more research might result in additional causal formulations, the temporal asymmetry (i.e., social background preceded course of study) makes reversal of the causal chain implausible.

Reliability of the various measures should be discussed. As Duncan (1969b) has noted in connection with American data, psychological variables, as generated from questionnaire data, are likely to have lower reliabilities than easier-to-measure socioeconomic variables. With these Turkish data, information on father's occupation, field of study, and place of work was gathered from both questionnaires and documentary sources. Reliability for these variables would seem to be quite high, while that of the job satisfaction variable would be expected to be somewhat lower. Split-half reliabilities averaged better than 0.85 for the job satisfaction indices (1956 and 1965). Reasonably good reliability for the

3. The social background data were dichotomized here. A relatively privileged background was scored as 1, while a less privileged background was coded as 0. Courses of study at the university were ranked as follows: political (scored as 3), finance (2), or administrative (1). The organizations were ranked according to Roos and Roos (1971): Ministry of Foreign Affairs (6); new and noncentral organizations (5); old organizations (4); Ministry of Finance (3); Ministry of Interior—central offices (2); Ministry of Interior—district governors and assistant district governors (1). Job satisfaction scores were summed from a number of indicators. Problems of measurement have been treated in detail in Roos and Roos (1971). Variances were generally similar for comparable variables across the respondent groups.

TABLE 2.2
DETERMINANTS OF JOB SATISFACTION: A MULTIPLE REPLICATION*

	Social Background (A) and Section at Faculty (B)	Social Background (A) and Organization (C)	Social Background (A) and Job Satisfaction (D)	Section at Faculty (B) and Organization (C)	Section at Faculty (B) and Job Satisfaction (D)	Organization (C) and Job Satisfaction (D)
1946–55 Graduates in 1956 ($N = 241$)	.34† (.47)	.36† (1.31)	01 (.02)	.86† (2.26)	.39† (.57)	.41† (.23)
1946–55 Graduates in 1965 ($N = 241$)	.34† (.47)	.27† (1.07)	.09 (.18)	.63† (1.82)	.28† (.41)	.34† (.17)
1958–61 Graduates in 1965 ($N = 69$)	.22 (.28)	.15‡ (.63)	.09 (.20)	.54†‡ (1.85)	.30† (.51)	.42† (.21)

*The product-moment correlation was used. Unstandardized regression coefficients are presented in parentheses. Similar results were obtained when 1965 nonrespondents were added to the data from 1946–1955 graduates in 1956 (total $N = 380$). Test-retest correlations for the job satisfaction variable have been reported in Roos and Roos (1971:251).

† This relationship was statistically significant at the .05 level.

‡ Differences between these correlations and those for the 1946–1955 graduates in 1956 were statistically significant at the .05 level.

FIGURE 2.12. Causal Formulation for Job Satisfaction Data

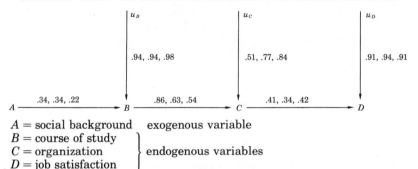

A = social background   exogenous variable
B = course of study ⎤
C = organization ⎬ endogenous variables
D = job satisfaction ⎦

u_B . . . u_D are residual variables influencing B . . . D, respectively.

The three numbers next to each arrow are the respective path coefficients for each path in the models for 1946–1955 graduates in 1956, 1946–1955 graduates in 1965, and 1958–1962 graduates in 1965.

Because no other paths were significantly different from zero, this simple chain model was selected.

job satisfaction item is suggested by the test-retest correlations which over a *nine-year* interval ranged from .49 to .63 for individuals remaining in the same types of positions in the same ministries. Because of the relatively simple nature of the causal model, the differential reliability of the variables should not cause difficulties.

Methodologically, the path analysis presented in figure 2.12 builds on the pioneering work by Duncan (1966) on causal modeling and occupational mobility. It represents an effort to apply techniques developed elsewhere to the study of elites and organizations. The panel and age-level data are most useful in permitting comparisons among the models and in assessing the generalizability of any single formulation. The availability of three sets of data provides a multiple replication of the original 1956 study. In a significant sense, the above causal model is the same for all groups, even though the correlations have been weakened between 1956 and 1965. This is particularly interesting because mean scores for the organizational and satisfaction variables changed markedly between 1956 and 1965. Because data were collected from new individuals in 1965, these findings cannot be explained by temporal erosion (i.e., the normal weakening of relationships over time [Campbell 1971:96–97]). Residual variables appear to have been more important in 1965 than in 1956. Since the variables presented above were not selected from a larger pool of hypothesized independent variables, cross-validation does not seem critical (McNemar 1969:207–09).

Historical analysis might suggest additional variables, although some of the determinants might be system-level factors difficult to incorporate with these data on individual administrators. Including other variables would affect relationships in the causal model. This would be particularly desirable because the variance explained in the dependent variable—job satisfaction—is rather small. The multiple correlation coefficients for job satisfaction $(r_{D \cdot ABC})$ are .44, .35, and .43, respectively, for the three groups of data. Finally, additional analysis might rely upon path coefficients, perhaps taking advantage of newer techniques for estimating path coefficients for unmeasured variables (Hauser and Goldberger 1971). System-level changes might be incorporated as "disturbances," or outside variables affecting measured variables.

There are several possibilities here. The use of cross-sectional data to assess causal relationships allows additional questions to be asked from the perspective of the pretest-posttest design. An event might change levels of variables or it might affect relationships. Changed relationships might differentially affect the plausibility of various causal formulations. If an event does affect relationships, causal modeling should illustrate various differences between treatment and control groups. Some of these possibilities are illustrated in figure 2.13. Such additional analysis would obviously be relevant for the Armor (1972) data on busing. Of course, data analyzed by means of causal modeling might also be approached from a pretest-posttest perspective. Goldberg's (1966) causal modeling of American voting behavior in 1956 could profitably be integrated with Pomper's (1972) pretest-posttest approach to 1956–1968 electoral data.

This analysis suggests that social change might be studied in terms of the levels of different variables, in terms of the weakening or strengthening of various relationships, and in terms of changes in over-all causal patterns. Methodological questions of constancy through time and equivalence among groups can be related to substantive concerns. Different causal models might prove plausible for some data sets and implausible for other related sets. These techniques might profitably be used in conjunction with panel analysis. Since less than 20 per cent of the variance in the satisfaction variable was explained in the model presented in figure 2.12, the possibilities for more "micro" approaches are obvious.

This macro-micro problem is of general significance. Questions

FIGURE 2.13. Causal Modeling with a Pretest-Posttest Design

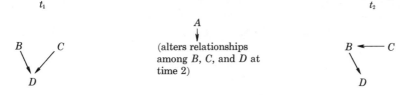

(alters relationships
among *B, C*, and *D* at
time 2)

Hypothetical Relationships in
Treatment Group in Pretest-Posttest Design

Hypothetical Relationships in
Control Group in Pretest-Posttest Design

A = event relevant for treatment group but not for control group
B, C, and *D* = variables measured at t_1 and t_2
For purposes of simplicity, residual variables are not included in this diagram.

concerning causality can often be better understood by subdividing the data along some theoretically relevant or policy-relevant dimension. In several studies, subdividing the sample led to different causal inferences from different sorts of settings (Crano, Kenny, and Campbell 1972; Roos and Roos 1971). Such controls provide a comparative perspective and contribute to theory-building. On the other hand, the variety of relationships prevailing in different settings may prove depressing. The generalizability of social science findings is threatened by such discoveries.

DESIGNS AND MEASURES

The use of multiple indicators and various data reduction techniques is well established in contemporary social science. Approaching a data set through multiple research designs is much less common but equally necessary. Such designs have been used before. Much of the information presented by Duvall and Welfling (chapter 3) in their quasi-experimental study has been handled in a different fashion by Morrison and Stevenson (1972; with Mitchell and Paden 1972).

This problem of design should be approached in a more self-conscious manner than it has thus far. In this paper, I attempt to do so by presenting some of the biases, problems, and opportunities associated with several different quasi-experimental designs. Some of the difficult research choices involved might be explained further.

Often there will be tradeoffs between a cross-sectional, multiple-measure approach and a stronger design or designs based on fewer measures. Caporaso's discussion of complementary operationalism in chapter 1 draws attention to the differences in "perspective and theoretical focus associated with each different method" of data collection. Other researchers have pointed out that "different indicators may explain different portions of the variance in the construct, having little common variance" (Sullivan 1971:334). If this does occur, looking at the empirical associations among indicators is an inappropriate tactic.

Although extremely high degrees of convergence among independent measures may be difficult to achieve, calculating the degree of association among these measures does provide one way to assess scoring methods. This technique has been highly developed through the use of multitrait, multimethod matrices (Campbell and Fiske 1959; Althauser, Heberlein, and Scott 1971). If relatively few measures are used, this approach to checking on validity is less useful. The choices are schematically presented in figure 2.14. If the investigator has a relatively high degree of confidence in his measure or measures, then more ambitious research and stronger designs are appropriate. On the other hand, greater attention to measurement is called for when considerable disagreement among measures is noted. Given limited resources, an increased emphasis on measurement may force recourse to a weaker design.

The inclinations of the author are toward choosing the stronger research design at the expense of multiple measures. Such inclinations have both theoretical and practical underpinnings. First of all, the perspective of complementary operationalism may suggest less of a stress on multiple indicators than that forwarded in the 1960s. Second, stronger research designs provide more checks on threats to reliability and validity than do weaker designs; the

FIGURE 2.14.   Overlap between Measures and Research
Strategies

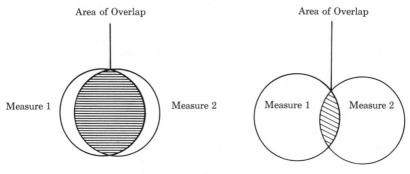

Area of Overlap

Area of Overlap

Measure 1   Measure 2   Measure 1   Measure 2

High Overlap between   Low Overlap between
Measures   Measures

Research Resources to   Research Resources to
Go into Stronger Design   Go into Dealing with
and Collection of Longi-   Measurement Problems
tudinal Data

over-all effort is improved by a stronger design. From a practical standpoint, additional measures and indicators can frequently be generated later, with or without additional funding. A researcher with a strong but complicated design may find comparatively few settings in which his data can be gathered. When such settings are available, substantial efforts at data collection would seem appropriate.

A related problem concerns choosing the research strategy which maximizes generalizability of the findings. Thus, in a comparative urban study, either cross-sectional or over-time tradeoffs between the number of research sites and over-all response rates are likely at any given level of resources. Przeworski and Teune's (1970) strategy of "most different systems" — maximizing variance in site selection and thereby in a number of presumably independent variables — is designed to achieve "generalizability." By choosing a relatively small number of very dissimilar sites, adequate variance in variables known or suspected to be significant may be obtained. In applying this logic to the problem of response rates, one becomes aware of the importance of minimizing nonresponse and gathering all available information on nonrespondents. As discussed earlier, respondents characteristically differ from non-respondents along a number of dimensions. A relatively high non-response rate makes the question of generalizability to entire populations particularly acute; the findings may not apply to re-spondents at one end of a number of continua. Nonresponse per-haps poses the main threat to the generalizability of most panel research, although collecting individual data from two or more time points does provide some checks on later nonresponse. Since considerable information about the respondents is generated at t_1, the characteristics of t_2 nonrespondents can be compared with those of respondents.

EVENTS AND VARIABLES

Some different conceptions of "reality" are implicit in the designs dealt with here. The designs differ markedly in their treatment of change. If perfect knowledge were available, the number of situations seen as truly discontinuous would undoubtedly diminish. But lacking such knowledge, it seems wise to conceptualize social phenomena in both continuous and discontinuous terms, using multiple replications and time-series data.

Questions of conceptualizing events and variables must be ex-plored further. A number of recent studies have dealt with the role of events in pretest-posttest fashion. Several examples of such research are outlined in table 2.3. In each case, additional analysis of the data according to panel analysis techniques would seem appropriate. Rather than a number of variables being di-rectly altered by a given event, a certain subset of variables might have been affected. Changes in one or more of these variables might then have caused a shift in other variables. Such analysis would contribute to an understanding of the causal dynamics of event-induced change.

The basic designs presented in this essay differ markedly in their treatment of intervening events. The following quasi-experimental designs have been compared:

$O_1 - O_2$ panel design (no inter-vening event)

$O_1 - R - O_2$ rotation design (event shuffles respondents)

$O_1 - X - O_2$ (+control groups)   pretest-posttest design
(event may affect re-
spondents)

In ongoing social systems it is often unclear which design is appropriate, and for this reason multiple research designs are recommended. Data analysis will aid in assessing the degree to which the assumptions implicit in each design have been met, thus providing strong clues as to the appropriate approach to the data. In the future, more advanced techniques for combining designs and analytical strategies may be helpful. Techniques of parameter estimation, whereby the effects of events can be compared with those of other variables, also would seem particularly useful. Path analysis and estimating path coefficients for unmeasured variables (impacts of events, perhaps) would seem most promising in this connection. Heise's (1970) model for panel data is suggestive here.

Certain dangers are connected with this "multiple design" strategy. In many situations data analysis might be conceived as a two-stage process. First of all, data are used to test the assumptions underlying one or more quasi-experimental designs. In the next stage, the design which has been selected is used to

TABLE 2.3

EVENTS AFFECTING INDIVIDUALS: SOME RESEARCH EXAMPLES

Specific Event	Effect on Individuals
American presidential campaign of 1964	"Increased voter consciousness of policy questions, and the later electoral persistence of group divisions" based on these questions (Pomper 1972:425).
French revolt in May of 1968	Voters "reexamine their habitual party preference in the light of underlying values" (Inglehart 1971: 992).
Decrease in financial resources of three Connecticut research and development organizations in late 1960s	"Greatest changes occurred in the way researchers perceived their jobs and their organizations rather than their self-perceptions or attitudes toward their work" (Hall and Mansfield 1971:533).

make substantive statements about the data. Measurement error and sampling error are likely to affect the researcher's choice among research designs. When the same data are used at both stages in the research process, the possibilities for substantively significant error are compounded.

There are several ways to deal with this problem. As McNemar (1969:209) has suggested, formulating one's model with a first sample and testing it with a second, independent sample provides a much less biased approach to data analysis. Advanced statistical procedures based on two-stage least squares regression models may be helpful (Johnston 1963); but, given an adequate N, the separate samples technique seems generally preferable. From another standpoint, explicit decision rules for choosing among quasi-experimental designs would help lessen "fudging" by the investigator in marginal cases. Further analytical and empirical research along these lines would be welcome.

Moving back and forth between different research tactics can help in the process of separating viable hypotheses and theories from those which are less realistic. Many phenomena can be conceptualized and researched in different ways. In the long run, this is likely to be valuable for methodology, theory-building, and policy analysis, but in the short run vigorous controversy seems inevitable. Cross-sectional studies characteristically analyze a wide range of variation in independent and dependent variables. When both theory and measurement techniques are relatively advanced, such studies may be particularly useful for parameter estimation (Johnston 1963:207). However, difficulties with collinearity, partial correlation, and so forth may well lead to results differing substantially from those obtained from time-series research. Frustration may be experienced; it probably is harder to get "interesting" (i.e., positive) results with well-conceived quasi-experiments than with other, weaker designs. The presence of numerous rival hypotheses is likely to result in the eventual rejection of many propositions generated by cross-sectional research. Such winnowing of hypotheses is, however, one viable strategy for advancing social science theory.

PART TWO

TESTING THEORY

In recent years there has been a renewed interest in developing social theory that is disciplined and undergirded by precise and sophisticated methodologies. The major obstacle to such an achievement has posed itself as a dilemma. On the one hand, the most powerful and abstract theories have weak methodological foundations and inadequate empirical grounding. Some variants of systems approaches, structural-functionalism, and rational models of coalition behavior share this weakness. On the other hand, strong methodological applications have been extensively made in areas where theory has been very weak.

In chapter 1 of this volume the issue was raised as to whether the quasi-experimental approach was conducive to building systematic theory. Although some limitations of a practical nature were conceded, the argument was put forward that there are no necessary obstacles to theoretical uses of this perspective. In this section the authors try to utilize a variety of quasi-experimental designs in evaluating their hypotheses.

In chapter 3, by Duvall and Welfling, the central analytical focus is on the process of political institutionalization. The authors are working solidly within a theoretical tradition generated by Samuel Huntington, Karl Deutsch, and S. N. Eisenstadt. After surveying the literature, they distill some of the salient hypotheses concerning the relationship between political institutionalization and three independent variables: the end of colonial rule, the occurrence of civil strife, and social mobilization. They then display some of the flexibility of the quasi-experimental perspective by tailoring individual research designs to the needs of each of their hypothesis tests. The "end of colonial rule" and "occurrence of civil strife" were both conceived as discrete events occurring within a one-year time interval. Thus a multiple, interrupted time-series design was chosen to assess these relationships. On the other hand, "social mobilization" was conceptualized as a continuous variable which is more appropriately researched through the application of a cross-lagged panel design. While the two designs differ in some significant respects, both allow one to probe relationships and search for causal dependencies.

Chapter 4 involves testing hypotheses about the systemic nature of relationships in the European Community. In contrast to the paper by Duvall and Welfling, Caporaso's article involves theory-testing at a more implicit, tentative, and preliminary level. Tests are exploratory and probative rather than decisive and conclusive. Starting from the position that regional integration is composed of two separate components, growth and coordination, he attempts to evaluate the extent to which the EEC possesses the latter.

To evaluate the extent to which system linkages have developed in the EEC, three quasi-experimental events were chosen: the formation of the EEC; the huge agricultural marathon of December, 1964, and the subsequent adoption of legislation concerning cereal prices; and the agricultural crisis of 1965–66. The reasons for these selections varied. The formation was chosen not primarily for theoretical reasons but because it was an important event whose impact on the institutionalization of interest group activity the author wanted to assess. The package deal in agriculture was chosen because agriculture is one of the more controversial, less technical, less functionally specific sectors of cooperation. The working hypothesis was that less differentiated sectors would be more closely involved with other sectors and that this lack of structural independence would result in spillover from agriculture to additional spheres of integrative activity. Finally, the agricultural crisis of 1965–66 was chosen not only because of the properties of the agricultural sector, but also because this was a negative experience for community integration. Since most of the change occurring in the EEC is in an upward direction, a predicted decrease in community integration provides a "riskier" hypothesis.

Chapter 5, by Michael Stohl, is within the linkage politics tradition. In this paper Stohl attempts to examine the relationship between domestic and international violence. Although a great deal of work has been done on the relationship between internal and external conflict, almost all of this has been of a cross-sectional nature. By testing out these relations over an extended body of time-series data, Stohl adds an important new dimension to the conflict literature.

For sources of theoretical insight, Stohl draws freely on the sociological writings of Max Weber, Karl Marx, Talcott Parsons, and Pitirim Sorokin, as well as on the more recent contributions to the theory of social conflict (e.g., Gurr 1970; Nardin 1971; and Jenkins 1969). Stohl synthesizes these diverse works and suggests specific hypotheses to be tested. Like the other authors in this section, he chooses the interrupted time-series design as the primary technique with which to evaluate his hypotheses. His findings should stimulate future exchanges among conflict and peace researchers.

Although none of the authors is engaged in constructing and testing formal, deductive theory, all are examining relationships and probing interdependencies at varying levels of abstraction. Finally, all are concerned with the process of winnowing involved in establishing knowledge claims, and each attempts to fulfill this goal methodologically by eliminating those hypotheses which are the major rivals to the hypothesis being tested.

3

DETERMINANTS OF POLITICAL INSTITUTIONALIZATION IN BLACK AFRICA

RAYMOND DUVALL

MARY WELFLING

The characteristic concern of "development" literature is with factors causing (or failing to cause) nation-states of the Third World to behave like "modern" polities. While a myriad of definitions of the concept of political development exists (Pye 1965; Diamant 1966) and a plethora of independent variables has been hypothesized to cause this ambiguous concept, it is possible to find a diversity of works which converge to provide a testable list of causal variables. For present purposes we are interested in hypotheses relating to the creation or building of viable political institutions capable of performing some set of functions for and in the social system (Eisenstadt 1962). In particular, we take as an interesting potential component of "development" the institutionalization of political party systems as primary two-way transmission belts between publics and government.[1] The research

This article is reprinted, in a slightly revised form, from *Comparative Political Studies,* Volume 5, number 4 (January, 1973), pp. 387–417, by permission of the publisher, Sage Publications, Inc.

1. A concern with political party systems is not a new one, since the development literature has focused on political parties more than on any other political institution with the possible exception of the bureaucracy. A primary function for political parties in new nations is an ability to further national integration (Zolberg 1963; Coleman and Rosberg 1964). Other functions attributed to parties include providing

reported here involves an attempt to assess the importance of hypothesized determinants of party system institutionalization in new nations.

An examination of literature sufficed to indicate that at least three social phenomena are consistently believed to be influential in affecting the building of viable political institutions. The first phenomenon is simply the end of colonial rule. A second is the occurrence of civil strife, and a third is rapid social mobilization and/or economic development. Each can be conceptualized as a form of strain on the political system, although disagreement emerges as to the impact of such strain on the building of political institutions.

SUBSTANTIVE HYPOTHESES

END OF COLONIAL RULE

The end of colonial rule is hypothesized to affect party system institutionalization through two primary mechanisms. First is a dissolution of unity. The achievement of the goal which has mobilized and unified a population is apt to result in a politically mobilized and active public without a unifying purpose. Such a situation is highly conducive to social conflict among ethnic or religious groups and in-fighting among political elites (Willner and Willner 1965; Wriggins 1961). The latter condition is held to be detrimental to party system institutionalization in failing to provide a stable environment for institutional development and in encouraging a focus on the ends of power and control at the expense of means of recruitment and political process. In this vein, Eisenstadt believes that

> possible cleavages with the elites in their pursuit of popular support may easily create conditions under which the elites may be unable to assure the initial institutionalization of

---

an ideology helping to maintain an adhesion to modernization policy, mobilizing support and participation, developing a national communications network, and recruiting and training leadership (Silvert 1965; Douglas 1963; Hess and Loewenberg 1964). The principal spokesman for the "primacy" of an institutionalized party system, however, is Samuel Huntington (1965, 1968), and it is through his work that much of the research reported here was stimulated.

political frameworks capable of absorption of change and may give way to regimes with a lower level of such ability. (1967:261)

The second mechanism is a dispersion of political resources (especially personnel) immediately after independence. During the colonial period those few members of the indigenous population who were enabled to acquire "modern" political skills tended to be concentrated in the party system. But the sudden emergence of new political roles (e.g., in the bureaucracy) siphoned off personnel from the party system, resulting in institutional weakness. In Huntington's terms, "[a] marked dispersion of resources means a decline in the overall level of political institutionalization" (1965:425). Similarly, Foltz asserts that "unless new sources of loyal organizational and administrative talents can be found immediately, the party's organization — and, therefore, the major link between the regime and the masses — is likely to be weakened" (1963:124). The argument is extended by Wallerstein (1966), who argues that Africa is experiencing a transition from one-party states to no-party states as party personnel become absorbed into the government bureaucracy and subsequently lose their dedication to party activities.

Because both mechanisms converge on an anticipated effect, theorists of political development generally hypothesize that national independence (i.e., the end of colonial rule) has had a debilitating effect on party system institutionalization. In particular, a testable hypothesis is that "[in] terms of institutional strength, . . . the new nations reached their peak of political development at the moment of independence" (Huntington 1965:408). Subsequently, political institutionalization is believed to have declined. But while the literature is virtually unanimous, the authors do not believe that this essentially descriptive (as opposed to theoretical) proposition is without plausible alternatives. On the contrary, it could well be argued that the colonial experience was and is the debilitating element, and that Third World nations are having some success in recovering from that experience by gradually building viable political institutions of national scope. The fact of independence and, hence, an ability to freely probe and experiment may encourage institutionalization. This apparent disagreement might be due solely to a focus on institutionalization as state on the one hand and institutionalization as process on the other. That is, all might agree that the immediate strain of independence lowered the level of apparent political institutionalization but, simultaneously, that a stimulating or threatening environment fosters a process of becoming more institutionalized (Buckley 1967:128–61). If such is the case, it becomes necessary to distinguish conceptually between state and process in assessing the impact of social strain on political development, a distinction which was suggested previously by Huntington (1965) but one which neither he nor others has actively pursued.

CIVIL STRIFE

The second phenomenon of interest is civil strife. Again, at least two conflicting arguments can be found. The first is that moderately shocking events in the environment of a system will stimulate adaptability and hence institutional strength on the part of that system. Representative is Huntington's statement that "fierce conflict or other serious challenges may transform organizations into institutions much more rapidly than normal circumstances" (1965:395). Also relevant is Popper's conclusion (1971) that internal war, if not too severe, can be a stimulant to political development. The alternative is that institutions develop most rapidly in secure, stable environments. In this vein, Weiner (1965:64) argues that processes of national integration will be stimulated by the learning of rules of social conduct, and national disintegration by the breaking of those rules. Similarly, Foltz (1963) maintains that nation-building can occur best in circumstances where elites feel secure. The latter (stable-environment) argument seems to be most consistent with the literature summarized in the discussion of national independence, where agreement is widespread that post-independence political "decay" is partially attributable to a dissolution of unity and concomitant civil conflict. On the other hand, the former argument seems potentially compatible with the notions suggested at the end of the discussion on independence. If the stability and constraint of the colonial period are seen as debilitating, the shocks and stimulation of independence and subsequent occurrences of civil strife may encourage the party system to develop institutional strength in linking publics and government. At any rate, the researcher should be encouraged if the relationships that obtain between strife and institutionalization are supportive of those involving national independence, since each is an environmental strain with potential effects on both the state and the process of institutionalization.

SOCIAL MOBILIZATION

The third variable of interest is social mobilization. Problematic is the multidimensional nature of the concept and concomitant disagreement about the relationships among dimensions. Deutsch, in his early exposition of mobilization (1961), argued that the concept is essentially unidimensional, with all of the components of mass media exposure, residential change, urbanization, occupational change, literacy, and wealth varying together synchronically and diachronically. For Deutsch (1961:501), the relationship between mobilization and political development is primarily dependent on the social composition of a nation, such that mobilization will increase political development in homogeneous societies while impeding it in heterogeneous nations.

Except for Deutsch, most authors select for consideration only one component of the more general concept of social mobilization. The two most commonly isolated are increased political awareness on the part of mobilized publics (generally due to increased education, literacy, and mass media exposure) and increased economic wealth. Virtually everyone who focuses on the former (Foltz 1963:126–28; Huntington 1965:415–17; MIT Study Group 1967:46) notes the debilitating potential of increased demands from publics for governmental performance and the destabilizing effect of new prospective elites seeking unavailable positions of authority. In terms of the party system as a primary channel between publics and government, increased awareness is expected to strain that channel to the point of overload with demands, and to foster circumvention of that channel by impatient elites. Both effects operate to cause a decrease in party system institutionalization.

Alternatively, it might be argued that increased awareness serves to counteract the problems of personnel diffusion noted in the discussion of national independence by creating new elites, thereby affording the party system a necessary resource base for institutionalization. In addition, increased awareness might be thought to facilitate communication by fostering a more widespread acceptance of shared symbols, desires, and expectations (Pye 1963; Schramm 1964; Lerner and Schramm 1967). In this respect, greater awareness could be hypothesized to promote party system institutionalization. In the same way, persons concerned with economic development have argued that increased wealth should mean increased governmental capability which, in turn, should imply a more effective, more institutional party system linking publics and government. Again, however, that might be countered by an argument to the effect that increased wealth serves to increase expectations and demands at an even faster rate, thereby weakening the ability of the party system to provide a satisfactory linkage (Olson 1963).

It is clear, then, that a set of partially contradictory plausible hypotheses exist regarding the relationship between social mobilization and party system institutionalization. That relationship might be entirely a function of some third variable, as Deutsch (1961) suggests, or it might be a complex of relationships such that strains become temporary overloads and reduce the state of institutionalization but simultaneously encourage a positive process of institutionalization as the party system adjusts to new demands and resources. We address ourselves to the several hypotheses relating social strain and the state or process of political institutionalization in an effort to empirically assess their validity.

THE DESIGN OF RESEARCH

Scholarly literature has given us a set of testable hypotheses involving the institutionalization of political party systems as a dependent variable. All are seen to have plausible alternatives if only in the form of direct opposites. Our intent is to assess the plausibility of each alternative as rigorously as possible and to eliminate as invalid those which the evidence shows are implausible. But our purpose does not end with selecting between two antithetical explanations. Rather, while offering tests, it is to be aware of and control for as many as possible of the nonobvious explanations of variations that occur in institutionalization. We find quasi-experimental designs most satisfactory for that task.

Three methodological strategies are generally enumerated for the purpose of achieving control: randomization, matching, and statistical manipulation (Lijphart 1971). Randomization lies at the heart of true experimentation and involves the random assignment of subjects, or cases, to groups for which at least one manipulated variable is assigned different values. As randomization can be used to obtain representativeness of populations from samples within certain specifiable margins of error, so can randomization provide control by establishing group equivalence on test-centered attributes within certain probabilities of error. Randomization is regarded as the optimal method of control, but is generally not utilizable by scientists dealing with natural social settings who involve themselves in ex post facto analysis of social givens. Since statistical manipulation requires an enumeration of variables to be controlled, its utility is entirely dependent on the intuitive facility of the analyst in selecting crucial variables. We believe that many potentially rival explanations are nonobvious, and, hence, are unlikely to be statistically controlled. The logic of matching remains as a useful strategy for scientists dealing with natural settings. Essentially it involves an attempt to achieve equivalence (hence, control) by a selection of entities, or packages of variables, which are highly similar in most attributes except those of theoretical interest. A major weakness of the method is a severely high probability that entities *cannot* be matched on *all* variables excepting theoretical ones. Some variables will always remain as plausible rival explanations, particularly if we recognize that systematic matching on some variables implies systematic unmatching on others. Entities, or systems of variables, may be matched spatially—the basis for "most similar systems" comparative research (Przeworski and Teune 1970)—or temporally—the basis for diachronic analysis of a single entity. We assume that, as a rule, there is less variation diachronically than synchronically; that is, most variables manifest less variance through time for a given entity than at one time for a set of entities. This assumption, together with the logic (though not the systematic rigor) of control through matching, is involved when a set of temporally based designs are labeled quasi-experimental.[2] The label is appropriate to the extent that temporally based observations on an entity approximate the equivalence expected from randomization.

In addition to affording strong controls for other variables in ascertaining the extent and form of relationships, quasi-experimental designs can enable a probing for causal direction. The latter is particularly important in the present research, since causal hypotheses are being tested and since the direction of causation is questionable. Specifically, while our hypotheses propose causal relationships from civil strife and social mobilization to political institutionalization, other studies suggest or propose the reverse. Gurr (1970:274) enumerates political institutionalization

2. The literature on quasi-experimental analysis has increased considerably in recent years, largely under the impetus of Donald Campbell. For discussions of the nature and logic of quasi-experimental designs see Campbell and Stanley (1963), Campbell and Ross (1968), Caporaso and Pelowski (1971), and Campbell (1963, chapter 6 of this volume).

as one variable acting to cause civil strife. Weiner (1965:63–64) argues that economic development is more apt to be a function of political institutionalization than vice versa. Supportive of Weiner are Hess and Loewenberg (1964), who maintain that Ethiopia is the least economically developed African nation in large part because it does not have an institutionalized political party system. Because of these rival hypotheses it is imperative to employ techniques that enable us to probe causality.

For each of the three variables discussed above, an appropriate quasi-experimental design was sought such that (1) a probable causal relationship could be detected, and (2) plausible rival explanations would be reduced to a minimum. In the case of national independence and civil strife, the "experimental" variable can be conceived as a relatively discrete event, occurring within a given time frame such as a year. In such cases forms of the interrupted time-series design are preferred. For our purposes a multiple time-series design with control group was selected. On the other hand, social mobilization, as measured through archives, is more appropriately conceptualized as a continuous process or general condition characterizing a society for longer periods of time. It necessitates a less "event-based" design. For that reason, reliance is made of the less "experimental," but adequate, cross-lagged panel correlation technique.

A design employing multiple replications (on different nations) of an experimentally interrupted time series offers strong controls for several potentially confounding factors, particularly when a control group is employed. The essence of the design is a comparison of several values of the dependent variable (institutionalization) prior to the "experimental" event, and values of the variable subsequent to the event. The researcher is interested in changes in the dependent variable that occur at the time of the experimental treatment. The quasi-experimentalist's task is to account for as many as possible of the potentially confounding factors that could cause the observed change, such as "random" fluctuation or "normal" development (see chapter 1). In addition, the multiple time-series design accounts for a major additional confounding factor (Campbell and Stanley 1963:55). The probability of changes in the dependent variable being caused by other events occurring at the same time as the experimental treatment is greatly reduced when the experiment is replicated many times and in different nations. To the extent that the experimental event affects many subjects simultaneously (as with national independence in Black Africa), the use of a roughly equivalent control nation renders implausible the potential impact of some other event possibly occurring in all nations in the sample at the same time.

The second design, based on cross-lagged correlations, is a valuable aid in reducing the equivocality of direction of causation in instances where two variables are related. The technique does not purport to explain the general, or synchronous, relationship between variables (perhaps a function of variables exogenous to the immediate consideration), but only to reveal increases or decreases in that relationship at lagged intervals. Thus the technique assumes that the causal hypothesis implies a temporal asymmetry. With that assumption, when the lagged correlation between cause at time T and effect at time $T+1$ is significantly larger than that between effect at time T and cause at time $T+1$ the plausibility of the test hypothesis is greatly increased. But where significant differences do obtain, the design is not as adequate as the multiple time-series design in controlling for plausible rival explanations (chapter 2; Crano et al. 1972). In particular, correlational asymmetry may be due to changes in the reliability or measurement specificity of one of the variables between time T and time $T+1$. Also, third variables, or what Roos (chapter 2) calls events, can always constitute plausible rival explanations in situations where their effect could operate through complex mechanisms creating asymmetries. Additionally, as Rozelle and Campbell (1969) point out, where synchronous cross-correlations are nonzero and where the lagged correlation from cause at T (A) to effect at $T+1$ (D) is larger than that from effect at T (B) to cause at $T+1$ (C), two hypotheses remain jointly plausible: that A increases D, *and* that B decreases C. For our purposes, a significantly larger lagged correlation between mobilization and institutionalization than vice versa could indicate either that mobilization increases institutionalization and/or that institutionalization decreases mobilization, assuming that the synchronous cross-correlations are positive. Of course, if one of the two remaining hypotheses is held implausible on a priori grounds the technique is of great value. In any case, utilization of cross-lagged techniques is appropriate only when synchronous correlations are stable or when they consistently follow some reasonable trend through time. The absence of stable correlations renders interpretation difficult (Crano et al. 1972). In general, the greater the number of time points employed in cross-lagged analysis across nations, the greater the controls for plausible rivals, since correlation instability is more apparent, changes in reliability and measurement specificity can more easily be detected (unless they change monotonically), and correlation attenuation can be better estimated. Roos suggests, on the basis of the literature on panel designs, that a more satisfactory strategy would compare cross-lagged partial correlations and path coefficients. While we recognize the legitimacy of Roos's argument in countering difficulties such as differential reliabilities, our analysis rests on a comparison of the cross-lagged zero-order correlations primarily because we assume that causal asymmetries that exist should reveal themselves (although perhaps not as precisely) through a comparison of the simple lagged correlations. That is, we do not expect partial correlations to reveal considerably different patterns than the zero-order correlations. Largely on the suggestion of Roos, we are currently engaged in a reanalysis of some of the data reported here in order to test this assumption. Thus, the reader should be cautioned that results reported here may not be definitive even for the research design and data set employed.

THE MEASUREMENT OF CONCEPTS

A first task in hypothesis-testing research is the selection of an appropriate sample. We selected a sample where cases are similar on as many variables as possible, except, of course, those employed in explanatory statements. The logic of that choice is consistent with the logic of quasi-experimentation; namely, "control for" as many as possible alternative explanations by obtaining near equivalence. The result is a loss in the ability to generalize — an increased threat to "external validity" — but a concomitant increase in the more essential "internal validity" (i.e., one is more confident that the relationships that obtain are not spurious).[3]

To satisfy the requirements of the "most similar systems" design we took as our sample the entire population of independent Black African nation-states excepting Ethiopia (which has had no party system), the white-ruled nations of southern Africa, contemporary colonies, and recently independent Swaziland, and

---

3. The terms "internal validity" and "external validity" are taken from Campbell and Stanley (1963). In particular, "most similar systems" designs constitute a threat to generalizability through the possibility of interaction between sample selection and the explanatory variable used — in our case, Black Africanness and social mobilization or civil strife.

Equatorial Guinea. The result is a sample of thirty-one party systems and/or nations.[4]

The hypotheses and designs enumerated above necessitate measurement of four concepts: political party system institutionalization, end of colonial rule, civil strife, and social mobilization. We can briefly suggest measurement strategies employed, but the reader is referred to fuller discussions elsewhere (Welfling 1971). The authors directed their attention most thoroughly to the concept of institutionalization, relying in most part, although with some modifications, on the efforts of others in measuring civil strife and social mobilization.

INSTITUTIONALIZATION

Inspired by, but not limited to, previous conceptions of institutionalization by sociologists and political scientists (Eisenstadt 1965; Buckley 1967; Huntington 1965, 1968; Polsby 1968; Keohane 1969; Hopkins 1969), we maintain that the concept is characterized by four essential components: (1) boundary—the notion that an institutionalized system is clearly differentiable from its environments; (2) stable patterns of interaction—elements of the system must manifest some degree of regularity or normalcy in their interrelationships, *and* the system as an entity must interact with its relevant environments with some degree of stability; (3) scope—an institutionalized system has a noticeable impact on its relevant environments; and (4) adaptability—an institutionalized system is constantly attempting to come into a better "fit" with its environments and is receptive to changes in those environments.

A set of indicators appropriate to the party system as the unit of analysis and reflective of one of the four defining characteristics of institutionalization were selected for measurement.[5] They are:

4. The thirty-one nations in the sample are: Botswana, Burundi, Cameroun, Central African Republic, Chad, Congo (Brazzaville), Dahomey, Gabon, Gambia, Ghana, Guinea, Ivory Coast, Kenya, Lesotho, Liberia, Malawi, Mali, Mauritania, Niger, Nigeria, Rwanda, Senegal, Sierra Leone, Somali Republic, Sudan, Tanzania, Togo, Uganda, Upper Volta, Zaire, and Zambia.

5. We adopt here the perspective of multiple as opposed to definitional operationism in an effort to produce a less errorful measure of institutionalization than any of the indicators taken separately. All indicators are recognized as *im*perfect approximations of the concept, but to the extent that the separate indicators share variation, that shared portion can be interpreted as a more valid measure of the concept (Campbell 1969a).

Boundary:

1. Elected nonparty personnel—the percentage of legislators elected as "independents." Indicative of the unambiguous nature of the party system as primary link between publics and government.

Stable interactions:

2. Party splits—the number of splits weighted by the size of parties affected. Overtly indicative of nonstable interactions.

3. Mergers—similar to splits.

4. Alliance creation and dissolution— similar to splits.

5. Different entities—the creation of new parties or the disappearance of existing parties except through splits, mergers, alliances, bannings, or declaration of legal single party. The number of parties affected weighted by size. Indicative of a required adjustment by continuing parties to different components in the system.

6. Name change of parties—similar to new parties, but a more inferential indicator of alterations in the modes and patterns of interaction.

7. Change in relative strength of parties—yearly percentage changes in legislative seats held. Indicative of externally determined alterations in roles of interacting components.

Scope:

8. Electoral participation—the percentage of voting age population voting in national elections. Indicative of the perceived importance of the party system as a "transmission belt."

Adaptability:

9. Electoral discrimination—the official postponement of elections, invalidation of elections, and/or use of single list. Indicative of the

> ruling party or coalition being unable to adapt to environmental changes.
> 10. Bannings—the number of bans weighted by the size of parties banned and by the effectiveness of each ban in eliminating the "threat" or viability of the party banned.
> 11. Arrests of party personnel—similar to bannings.
> 12. Declaration of legal single party—inferential indicator of lack of adaptability.

Each indicator was scored for each year of the twenty-five-year period from 1946 to 1970 in which a party system of indigenous parties (or party) existed.[6] The result is a data cube of twelve indicators [7] for thirty-one countries and up to twenty-five years. A principal components analysis of the indicators for the thirty-one party systems aggregated for all years reveals an essentially unidimensional pattern. A first principal component accounts for more than 50 per cent of the variation among a reduced set of ten indicators. The concept of institutionalization as applied to party systems seems adequately indexed, then, by these indicators. In order to weight the indicators equally and to facilitate a handling of missing data when the indicators are treated yearly, we utilized a technique developed by Janda (1971), and produced as our yearly measure of institutionalization a standard score (Z) across the set of indicators. These Z scores were produced in two ways: (1) over all countries over all year, and (2) over all years for each country taken separately. Scores for the aggregate period obtained with the Z-technique correlate with the factor scores from the previous analysis at .97, hence, are indexing essentially the same phenomenon.

6. The scoring of indicators was done as archival research. A great variety of sources were searched, varying with the country in question. Utilized for all countries were *Africa Diary; Africa Report; Africa Research Bulletin;* Reuters (1967); and Segal (1961). In addition, hundreds of country-specific works were examined.
7. An independently coded judgmental variable, National Orientation of the Party System, was taken as indicative of scope, in that it focused on the relative basis of the system in regional/ethnic groupings as opposed to entire national populations. This variable was not included in the analysis reported here since it was not coded yearly but only at the time of independence. It correlated with our summary index of institutionalization at approximately .80.

END OF COLONIAL RULE

Information on dates of national "independence" is readily available. Singularly problematic is the selection of the appropriate event for each country such that a criterion of equivalence is satisfied. In making a selection we attempted to answer the question, "At what point did indigenous political personnel feel themselves independent or relatively unconstrained in comparison to prior experience?" Answers seemed in most part dependent on an element of "surprise." For French and Belgian colonies, where colonial socialization was essentially in terms of participation in the singular "mother empire," the granting of internal self-rule prior to independence constituted the most significant ending of colonial power (although, of course, Belgian colonies had either no period of self-rule or a very brief one). For British colonies, on the other hand, independence was anticipated for longer times. Devolution of colonial power was more gradual, and ended most significantly at the actual granting of independence. Thus we used these two events, self-rule and independence, as equivalent for the two sets of colonies as indicated.[8]

CIVIL STRIFE

A listing of strife events for all Black African countries was made available to us by the African National Integration Project. A classification of events by type is reported in the book *Black Africa* (Morrison et al. 1972), for the shorter period of 1955 to 1969 or independence to 1969. For our purposes, events of the following types were included: demonstrations, strikes, riots, declarations of emergency, interethnic violence, irredentist activity, rebellions, and civil wars. We did not create a scale. Rather, the number of events of each type for each year were simply counted. Years with greater than a threshold number of events were utilized as "experimental" years.

SOCIAL MOBILIZATION

Again, we utilized data from the African National Integration

8. Developments in the UN Trust Territories of Cameroun (East) and Togo were influenced largely by the French presence, and thus self-rule is considered the more significant event. Self-rule was also used as the significant event in the Italian Trust Territory of Somali.

Project. We divided a set of indicators into three five-year periods: 1955–59, 1960–64, and 1965–69. A separate index of social mobilization was created for each five-year period by generating factor scores for the first principal component. For each of the three periods the first principal component accounted for approximately 50 per cent of the variance in the set of indicators (the values are 56 per cent, 46 per cent, and 53 per cent, respectively) so that, while the separate scores include somewhat different indicators,[9] there is some support for Deutsch's unidimensionality hypothesis, and the resulting index seems to adequately represent the concept of social mobilization. In addition, however, we utilized in our cross-lagged correlation analysis a set of indicators that were measured individually for at least two of our three time periods. This analysis involving separate indicators was included to test explicitly the propositions enumerated above hypothesizing a potentially different impact of political awareness and economic development on political institutionalization.

RESULTS OF ANALYSIS

Because the interrupted time-series design is dependent on a reliable estimate of trend in a series, and because very short time series are highly unreliable, we included only countries that have had a minimum of six years of indigenous party system experience both prior and subsequent to the "independence" occasion. Eighteen of the thirty-one countries satisfied this criterion, four involving independence and fourteen involving internal self-rule. The mean time-series length for the eighteen countries was nine pretest and eleven posttest observations.

Table 3.1 lists the countries and test design information, and gives the values which exceed tabled values at the .05 level of significance for the two tests of significance used. The tests of significance employed are given full explication in Sween and Campbell (1965). The first is concerned with a shift in the level of

9. The indicators included in the principal components analyses were: 1955–59 — radios, urbanization, vehicles; 1960–64 — radios, urbanization, wage earners, coal consumption, primary education enrollment, females in primary schools, adult population with fourth-year education, adult population with education beyond secondary school, size of government budget; 1965–69 — radios, urbanization, automobiles, newspapers, literacy, secondary education, primary education, number of doctors, GNP, and agricultural labor. All, of course, were standardized (e.g., per capita) as reported in Morrison et al. (1972).

TABLE 3.1
THE IMPACT OF THE END OF COLONIAL RULE ON
PARTY SYSTEM INSTITUTIONALIZATION *

Country	Occasion	Observations †	Difference in Slope (F)	Difference in Level (t) ‡	Autocorrelated Error §
Cameroun	Self-rule	1950–70 (7–14)	8.76		.02
Central African Republic	Self-rule	1947–65 (10–9)			−.28
Chad	Self-rule	1947–70 (10–14)		3.09	−.25
Congo (Brazzaville)	Self-rule	1947–63 (10–7)			−.07
Dahomey	Self-rule	1947–65 (10–9)			.13
Gabon	Self-rule	1949–70 (8–14)			.01
Ghana	Independence	1950–65 (8–8)		2.27	−.71
Ivory Coast	Self-rule	1947–70 (10–14)			.07
Malawi	Independence	1959–70 (6–6)	6.59		−.65
Mali	Self-rule	1947–68 (10–12)		2.35	.20
Mauritania	Self-rule	1949–70 (8–14)		3.24	.03
Niger	Self-rule	1949–70 (8–14)			−.15
Senegal	Self-rule	1947–70 (10–14)		2.75	.04
Somali	Self-rule	1947–69 (11–12)		3.10	.04
Togo	Self-rule	1947–66 (11–9)		2.45	−.28
Uganda	Independence	1956–70 (7–8)			−.21
Upper Volta	Self-rule	1949–65 (8–9)	5.11		−.43
Zambia	Independence	1959–70 (6–6)	6.09		−.19
Liberia	Self-rule	1946–70 (11–14)			−.33
Liberia	Independence	1946–70 (15–10)			−.15

* Yearly scores are standardized separately for each country.

† Numbers in parentheses refer to pretest and posttest observations.

‡ Double extrapolation to experimental intercept.

§ Mean autocorrelation for pre- and posttest series separately.

the time series, and reflects a change in the average, or mean, condition of institutionalization. Its significance is determined using a t statistic with $N - 4$ degrees of freedom, where N is the number of observations, or years, in the time series. The numerator of t is the difference between two values of institutionalization predicted for the year of experimental treatment,[10] and the two values are obtained by extrapolation from the best linear description [11] (i.e., the least-squares line) of each of the pretest and posttest series separately. The denominator is a function of variability in the two partial series. The second significance test is designed to reflect a change in the trend of becoming more or less institutionalized. It is given as an F statistic based on the difference between the slopes of the pretest and posttest least-squares lines relative to variability in each of the partial series. The F is evaluated with 1, $N - 4$ degrees of freedom. In addition, table 3.1 gives the mean of the first (lag one) autocorrelation coefficients for deviations about pretest and posttest trends. While no explicit rule exists for evaluating the effect of autocorrelated error on trend estimates, or, hence, tests of significance based on those trend estimates, the autocorrelation coefficients can be taken as some guide to the tentativity of results.

An examination of table 3.1 reveals that it is probably a safe conclusion that the end of colonial rule has had a significant impact on the institutionalization of party systems. Eleven of the eighteen experimental replications produce significant differences on one of our tests of significance. Over 30 per cent of the possible tests of significance (eleven of thirty-six) are significant beyond the .05 level, and autocorrelated error seems to be potentially confounding in only three cases (Ghana, Malawi, and Upper Volta). Finally, and importantly, two control cases (both Liberia) reveal no sig-

10. The predicted values are actually based on a mid-year point. Extrapolations are made from the two partial series to a point halfway between the last pretest observation and the first posttest observation.

11. Both of the tests of significance employed are based on an assumption that the least-squares lines are, indeed, best linear unbiased estimates of trend. This may be a dangerous assumption in some cases where autocorrelation of residuals about trend is significant, since such autocorrelation violates the assumption of the classical linear regression model that error terms are independent and heteroscedastic. But while there may be some danger, we have performed the analysis with classical regression estimates of trend for two reasons: the trend estimates can be shown to be unbiased, if not best; and reestimating each trend on the basis of a covariance matrix of error terms (generalized linear regression) would have necessitated a great deal of computational effort with very little expected difference in test results.

nificant differences in institutionalization around 1957 or 1960, the modal years for granting of internal self-rule and national independence among the eighteen test cases.

The control case reduces the plausibility of "history" (i.e., other events that affected all of Africa in those years) as a rival explanation. Similarly, other common plausible rivals are controlled in this design as indicated in the discussion above (e.g., the plausibility of "maturation" or normal development as the cause is reduced by the multiplicity of replications, many experiencing the experimental treatment at different points in their "life"). One explanation, however, remains plausible and should be considered. It is "instrumentation," where changes in the measurement process at the time of experimental treatment could account for the observed differences. In this case "instrumentation" is plausible since certain of our indicators could only be scored for post-"independence" years (e.g., creation of a legal single party). Since our score is a *mean* standard score across all indicators, however, the inclusion of a new indicator per se should not affect the summary index. That is, the instrument was constructed so that the number of indicators for which there was information for any year should not affect the final score. "Instrumentation" as an explanation, then, while plausible, seems improbable. The probability is high that the end of colonial rule is a factor with real impact on the development of political institutions.

If we do conclude that "independence" has caused a significant difference in political institutionalization, we have only accomplished half of the task. Table 3.1 suggests that we are dealing with a variety of effects. Seven countries have experienced a jump in the level of institutionalization as indicated by a great difference in experimental intercept. None of those seven exhibits, in the post-independence period, a trend (slope) that is significantly different from its pre-independence trend (slope). Alternatively, four countries exhibit relatively continuous, but "jointed," time series, in the sense that trend is highly different for the two periods, but intercept is not different.

Table 3.2 provides information required to test the hypothesis developed above, namely, that a process of institutionalization through the colonial period has been significantly disrupted and replaced by a marked process of "decay" since "independence." The table quickly reveals that several patterns exist, not simply the hypothesized one. Of the eleven countries in which significant results obtained, only two, Somali and Zambia, manifest "decay" *both* in terms of general condition (mean level of the series) and

TABLE 3.2
Forms of Change in Institutionalization at "Independence" *

Country	Mean of Pre-Independence Yearly Scores	Mean of Post-Independence Yearly Scores	Pre-Independence Slope	Post-Independence Slope	Independence Year Extrapolated from Pre-Independence Series	Independence Year Extrapolated from Post-Independence Series
Cameroun	−.09	+.02	−.161	+.021	−.65	−.12
Chad	+.10	−.06	+.000	+.062	+.11	−.49
Ghana	−.06	+.08	−.104	−.019	−.48	+.15
Malawi	−.21	+.11	+.111	−.036	+.13	+.22
Mali	+.06	−.06	+.006	+.088	+.09	−.58
Mauritania	+.21	−.10	+.003	+.057	+.22	−.50
Senegal	+.12	−.07	+.026	+.053	+.25	−.44
Somali	+.27	−.14	+.048	+.037	+.53	−.36
Togo	+.20	−.14	+.002	+.085	+.21	−.53
Upper Volta	+.05	−.05	−.073	+.106	−.24	−.53
Zambia	−.01	−.02	+.295	−.153	+.87	+.44

*Positive scores represent higher institutionalization and negative scores represent lower institutionalization. Scores cannot be compared across countries since they are created as country-specific mean standard scores. Differences that are significant according to our two tests are underlined with a solid line; shifts in the level of a series corresponding to significant differences in experimental intercept are indicated by a broken line.

trend (slope). On the other hand, only Cameroun and Ghana capitalized on independence such that institutionalization is (was, for Ghana) *both* greater on the average and increasing in trend since the end of colonial rule. The remaining seven countries exhibit mixed patterns. One of these, Malawi, increased its level of institutionalization following independence, but during the post-independence period the trend has been in an uninstitutionalizing direction. It is highly possible, however, that the changes observed are spurious in the sense that they are based on so little variance on the indicators. Malawi's party system might better be characterized as exhibiting no real process. The other six countries — Chad, Mali, Mauritania, Senegal, Togo, and Upper Volta — are characterized by a sudden and severe burst of uninstitutional party activity immediately subsequent to self-rule, but through time that activity has decreased in magnitude such that the trend has been in an institutionalizing direction.[12] In these six cases we find a brief period of uninstitutional activity, followed by a consolidation of single-party rule (though not necessarily the creation of a single-party state) after which party activity stabilizes somewhat.

Consider the case of Chad, which is typical not only of these six cases but also of many of our party systems, including some that did not exhibit significant results. Prior to self-rule Chad experienced a series of uninstitutional events (primarily changes in legislative strength, the appearance of new entities, and changing patterns of interaction), but these instances provided no clear trend or process of institutionalization. Immediately following self-rule (1957–59), however, we find a jump in uninstitutional activity, with an increase in the representation of independents, the splitting of the previously major party (AST), a continual shifting of alliances among the parties, and a rapid alternation in legislative strength between the AST and Tombalbaye's PPT, which changed its status from minor party to the majority and ruling party. In the next period (1960–61) uninstitutional activity continued with

12. It should be pointed out that five of the eleven party systems which demonstrated significant results have been ended by coups: Ghana, Mali, Somali, Togo, and Upper Volta. Four of these five had lower levels of institutionalization after "independence," and the fifth had an uninstitutionalizing trend after "independence," although this trend was less negative than that for the pre-independence period. One might be tempted to make inferences about the relative importance of the two aspects of change in predicting to this ultimate form of party-system "decay," at least in the short run. But the validity of such inferences is dependent on more explicit tests than those offered here.

the attempts of the opposition to join forces in the PNA, abortive attempts of the PPT and PNA to ally in a National Union, and the first arrests of opposition leaders by the ruling PPT. In the period 1962–63, the party system seems to be concretizing while the PPT consolidated its power through arrests, bannings, and the creation of a single-party state. From 1964 through 1970, however, Chad's party system entered a period of quiescence with relatively few indications of uninstitutionalization (arrests in 1965 and 1967 and electoral discrimination in 1969), and, in spite of the continued civil war, the release of major opposition leaders and their incorporation into the PPT in 1971 suggests that this period of party system stabilization may be continuing in the 1970s beyond our period of analysis.

These various results point to the need for a clear distinction between the static condition of institutionalization (level) and the process of institutionalization over time (slope). Our analysis indicates that institutionalization condition and process are indeed different phenomena—those party systems which shift in level tend not to shift in slope, and vice versa. Moreover, for the seven countries in which a significant change in the level of institutionalization obtained, a clear pattern (six of seven) emerged of decreases in institutionalization subsequent to "independence," but for the four countries in which a significant change in trend obtained, the division was equal, with half exhibiting a process of "decay" and half manifesting a process of becoming increasingly institutionalized. The obvious conclusion is that simplistic statements to the effect that new nations have experienced political "decay" since "independence" are unfounded, particularly when process rather than state is implied. A good deal of variation exists in the real impact that "independence" has had. Future research will need to move to other system-level variables in order to explain the differences in form that we have ascertained here. Places to begin would be with those variables for which the event of "independence" is taken to be a surrogate. The variables most often suggested are the breakdown of national unity and/or the scarcity of political resources. The latter is, to our knowledge, not yet adequately measured except very indirectly through indicators of social mobilization. As previously mentioned, however, it is controversial whether that concept more nearly implies increased demands on existing resources or increased resources. Hints on the breakdown of unity, however, can be obtained directly from an examination of civil strife.

THE OCCURRENCE OF CIVIL STRIFE

As with "independence," we established certain criteria for including countries in the civil strife quasi-experimental analysis. Again we set six years as a minimum on which to base a pretest and/or posttest trend in institutionalization. In this case that criterion translated into including countries with at least six years (starting in 1954 at the earliest, since strife data was available only from 1955) in succession without strife, followed by at least one year of strife, and at least six more years (1) without strife, (2) with intermittent strife, or (3) with constant strife. This criterion is essentially that of a sharp upward jump in the level of strife either as an isolated occurrence or as the beginning of a period of conflict. In addition, countries were excluded if the prestrife years corresponded very nearly to the pre-"independence" period for the above analysis, since an overlapping of experimental events would render interpretation impossible as to which event was causing the observed effect. Finally, strife events were included only if greater than a certain "magnitude," since single demonstrations or strikes in a year are not expected to have any significant or lasting effect on party-system institutionalization. The result of this winnowing was a set of seven countries, and a total of seven quasi-experiments. Table 3.3 presents the relevant information regarding the impact of civil strife in the seven countries which met our selective criteria, and is interpreted in the same way as was table 3.1.

On first glance the results again seem somewhat ambiguous, with four of seven experiments yielding significant effects on the process of institutionalization, and six of fourteen tests of significance (43 per cent) producing values near or beyond the .05 level of random probability. But closer examination of the experimental events column of table 3.3 suggests that the relationship may be less uncertain. The four countries where clearly significant results are obtained are characterized by strife of relatively great intensity and extensive duration (i.e., the pretest and posttest series correspond respectively to times of relative internal tranquillity and heavy civil strife). Alternatively, the three countries in which effects were negligible or questionable were those experiencing isolated "bursts" of strife of various magnitudes. Thus, while there seems to be evidence sufficient to reject the hypothesis that isolated occurrences of civil strife contribute to either political "decay" or institutionalization, the real effect of more sustained strife remains highly plausible.

The multiple replication of the design renders improbable the

TABLE 3.3

The Impact of Civil Strife on Institutionalization*

Country	Division of Series (all begin in 1954)	Difference in Slope (F)	Difference in Level (t)†	Autocorrelated Error‡	Experimental Event
Chad	9–8	9.62		–.39	One riot and one declaration of emergency followed by armed rebellion and civil war
Gabon	11–6			.00	One riot and one strike and one declaration of emergency (in one year) followed by six years of peace
Liberia	8–9	12.25		–.25	One riot and one strike (in one year) followed by intermittent peace and turmoil
Mali	9–6			.44	Two-year armed rebellion followed by six years of peace
Mauritania	10–7	7.26	3.08	–.03	One strike and one declaration of emergency followed by three years of interethnic feuds, combined with riots, a demonstration, and a strike
Senegal	8–9	6.17		–.26	One riot, one strike, and one declaration of emergency followed almost every year by riots and strikes
Somali	9–7		2.07§	–.14	Two riots and armed rebellion combined in one year followed by seven years of peace

*The tests are based on the series of yearly mean standard scores produced over all countries.

†Double extrapolation to experimental intercept.

‡Mean autocorrelation for pretest and posttest series separately. The autocorrelation coefficients are not the same as those found in table 3.1, since a different series is employed for each country. Here we are concerned with impact on institutionalization relative to all years for all countries. For independence we were concerned with country-specific effects.

§Nearly significant at the .05 level.

hypothesis that severe civil strife has *no* impact. The tests of significance over the set of seven experiments are *too* different from that expected by chance. Similarly, the experimental treatment is introduced in different years for the four (or five) significant-effects countries. On these grounds, "history" (other simultaneous events) and "maturation" (normal developmental process) are rendered implausible. One explanation that remains uncontrolled here because of the elaborate criteria utilized in sampling is that of "statistical regression." The principle here, which could operate as a confounding effect, is that experimental groups are sometimes selected on the basis of their extreme scores, which a simple probability model predicts will have a strong tendency to move toward an average (mean) value on subsequent observations. If the five countries (Chad, Liberia, Mauritania, Senegal, Somali) were characterized by extreme scores on institutionalization (in either direction), one would predict a shift in their scores toward the mean. And if the experimental variable, occurrences of severe civil strife, is the *result* of extreme institutionalization (again, in either direction), its occurrence might be expected to be nearly coterminous with the shift due to the regression effect.

Table 3.4 provides some potential insights into this possible problem, as well as information on the forms of the relationships obtained. Once again, effects on the process of institutionalization

TABLE 3.4
FORMS OF CHANGE IN INSTITUTIONALIZATION WITH STRIFE *

Country	Mean of Prestrife Yearly Scores	Mean of Poststrife Yearly Scores	Prestrife Slope	Poststrife Slope
Chad	−.13	−.07	−.101	+.027
Liberia	+.35	+.37	+.027	−.008
Mauritania	+.26	+.01	+.013	−.029
Senegal	+.11	+.12	−.036	+.007
Somali	−.01	+.17	−.041	−.044
Gabon	+.21	+.26	+.027	−.054
Mali	−.06	+.06	−.026	−.003

* Positive scores represent higher institutionalization, and negative scores represent lower institutionalization. Scores are based on all years for all countries, and thus are comparable across countries. Significant differences are underlined with a solid line; shifts in series level associated with significant differences in experimental intercept are indicated with a dotted line.

are inconsistent. For Chad and Senegal, the outbreak of strife served to reverse a process of "decay," but for Liberia and Mauritania a process of "decay" obtained subsequent to strife. Where a significant impact on the level of institutionalization was observed, results were also divided. In Mauritania, the period of civil strife was characterized by lower levels of political institutionalization (more uninstitutional activity in the party system), while in Somali the reverse was the case. But it seems potentially important to note that in every case of significant change associated with strife except Somali, the change was toward (and often beyond) the mean, which is near zero in the Z-score technique used. That this shift is not merely due to statistical regression effects, however, seems apparent from the fact that neither significant changes nor shifts toward the mean are always associated with the most extreme scores. Thus Liberia, with the most extreme level of pre-strife institutionalization, increased that position subsequent to strife. What does seem to be the case is that where strife and political institutionalization are significantly related, uninstitutional party-system activity increases during and subsequent to civil strife in countries previously marked by low levels of such activity; but in countries where party-system "decay" was occurring prior to strife, the latter is associated with increasing institutionalization. If one were given to interpretive inferences, it might be ventured that highly institutionalized (or overinstitutionalized) systems "fall apart" under stress, while disorganized, uninstitutionalized systems adapt, pull together, or increase organization under stress. The safer, and less grand, conclusion is simply that prolonged and intense strife is associated with changes in political institutionalization, that those changes are more apt to be in terms of process than state, or level, and that the causal relationship underlying the association is still questionable in that the "experimental" variable (strife) *could* itself be a function of extreme scores on the dependent variable (institutionalization),[13] which would normally be expected to regress toward the mean.

SOCIAL MOBILIZATION

In selecting countries for inclusion in the cross-lagged panel correlation analysis involving social mobilization, our only cri-

13. It is interesting that the "regression" explanation operates at "extremes" of both directions. Thus Liberia and Mauritania are perhaps overinstitutionalized prior to civil strife, while Chad is quite uninstitutionalized. The implication is that civil strife may be a product of either type of extreme.

TABLE 3.5
SYNCHRONOUS CORRELATIONS FOR SOCIAL MOBILIZATION AND TWO MEASURES OF INSTITUTIONALIZATION OVER THREE TIME PERIODS AND TWO SAMPLES *

	Time Period	Social Mobilization Index †	Radios per Capita	Change in Radios per Capita	Per Cent Urban	Change in Per Cent Urban	Change in GNP	Secondary School Enrollment	Primary School Enrollment
I) Mean Level of Institutionalization									
Sample 1	1	+.21	+.26	−.05	+.24	+.17	−.28	−.09	
	2	+.32	+.48	−.17	+.02	+.31	+.26	+.12	
Sample 2	2	+.37	+.52	+.20	+.15		+.25		+.06
	3	+.41	+.55	+.27	+.41		−.02		+.15
II) Trend in Institutionalization‡									
Sample 1	1	+.55	+.48	−.02	+.27	+.10	−.06	−.37	
	2	−.21	−.27	+.42	−.13	−.13	−.11	+.12	
Sample 2	2	−.13	−.41	−.34	−.09		−.06		+.17
	3	−.29	+.06	+.15	−.35		−.12		−.04

*Institutionalization scores used were those generated by standardizing across all years for all countries.
† Factor scores from separate principal components analyses.
‡ Slope of least-squares line for five-year period.

terion was that the country have a functioning party system throughout at least two of the three five-year periods: 1955–59, 1960–64, and 1965–69. This criterion yielded eighteen countries for the first two time periods, and seventeen for the last two. Ten countries were common to both sets.[14] An advantage of using largely different samples is the same as that for the somewhat analogous split-halves technique in gauging reliability. If relationships over the two samples are not similar and stable, significant correlations can probably be dismissed as random and, hence, spurious in the truest sense.

This criterion of relational stability reduces the need to reproduce all of the cross-lagged correlation results. Table 3.5 provides the synchronous cross-correlations between each of our eight measures of social mobilization and two measures of institutionalization, the general level (or mean value) for the five-year period, and trend (or slope) during those same years.[15] Two variables, radios per capita and the multiple-indicator index, manifest strong and relatively stable relationships through time with the general level of institutionalization. In both cases the relationship is such that high levels of social mobilization are associated with high degrees of political institutionalization. Three other variables— urbanization, change in urbanization, and primary school enrollment—are probably worthy of examination in relation to mean level of institutionalization in spite of less stability and/or weaker correlations. No variable relates consistently *and* strongly with our trend measure (or short-term processes). Only change in GNP per capita exhibits consistent relationships, but these are weak. All other variables are deleted from subsequent analysis on the grounds that too many rivals and confounding explanations are left uncontrolled in cases of unstable synchronous correlation.

The cross-lagged correlation results for the six mobilization variables of interest appear in figure 3.1. In interpreting the results, it should be reiterated that the method makes no assumption about the operation or lack of operation of exogenous variables which may be the more fundamental causes of the general synchronous correlations that obtain. Rather, the focus is on the asym-

14. The ten of three periods' duration are Cameroun, Chad, Gabon, Gambia, Ivory Coast, Liberia, Mauritania, Niger, Senegal, and Somali. Limited to the first two periods were Central African Republic, Dahomey, Ghana, Mali, Nigeria, Sierra Leone, Togo, and Upper Volta. The seven used only for periods two and three were Guinea, Kenya, Malawi, Rwanda, Tanzania, Uganda, and Zambia.

15. The slope variable is, in this analysis, probably not terribly reliable, since five observations provide an unstable data set for estimating a regression equation. In that respect, the correlations with mean value are perhaps more interesting.

metry of cross-lagged correlations. If one of the two variables is relatively more a cause and the other an effect, the lagged correlation from cause to effect is expected to be larger than that from effect to cause. We look, then, at the differences between lagged correlations in order to test hypotheses of causality. A test of significance is used to assess the difference. We use the t test, with

FIGURE 3.1. Cross-lagged Panel Correlations Involving Social Mobilization Indicators and Political Institutionalization

A) Index of Social Mobilization

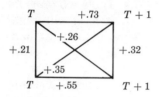

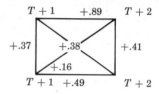

Mean level of institutionalization
$t = .27$

$t = .64$

B) Radios per Capita

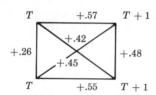

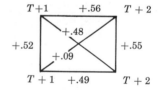

Mean level of institutionalization
$t = .10$

$t = 1.17$

C) Percent Urban

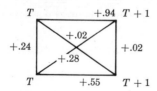

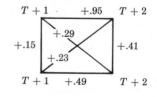

Mean level of institutionalization
$t = .72$

$t = .17$

FIGURE 3.1 *(Continued)*

D)   Change in Percent Urban

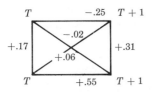

Mean level of institutionalization
$t = .22$

E)   Primary School Enrollment

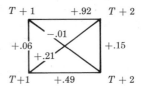

Mean level of institutionalization
$t = .60$

F)   Change in GNP

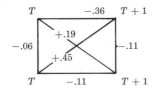

Trend in institutionalization
$t = .80$

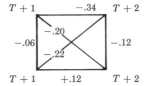

$t = .06$

Horizontal lines indicate the correlation of a variable with itself at two successive time intervals. Vertical lines indicate synchronous, bivariate correlations. Diagonal lines indicate cross-lagged correlations. Social mobilization indicators are always given as the top variable, and institutionalization appears on the bottom. The symbols at the corners of each figure indicate time intervals as follows: T corresponds to 1955–59; $T+1$ corresponds to 1960–64; and $T+2$ corresponds to 1965–69.

the z' transformation recommended by Fisher and given in Peters and Van Woorhis (1940), although we realize the somewhat conservative bias of that test. The t values appear with each subtable.

Of ten panels, *no t* value reached a level of minimal significance. It clearly cannot be concluded that an asymmetric causal relationship exists between social mobilization and political institutionalization, at least over five-year periods. At least five possibilities remain: (1) a symmetric, or simultaneous, relationship does exist such that each "causes" the other (plausible at least in terms of the composite index and ratios per capita since synchronous correlations for them are large); (2) both variables are "co-symptoms" of an exogenous variable, or simply unrelated (the latter is particularly plausible in the light of the generally low correlations);

(3) five-year periods are inappropriate time slices for uncovering the causal linkages operating; (4) the sample manifested so little effective variance on the "independent" variable (social mobilization indicators) that all we obtained was "noise" on a more pervasive and general relationship; or (5) some variable such as Deutsch's national homogeneity is operating, which, if controlled, would reveal a real relationship (that is, two or more populations of nations, each with relationships of a different form, might be included here as a single population, thus masking the relationships). In light of the large volume of literature on the relationship between economic development, or social mobilization, and political development, further empirical investigation is certainly in order, particularly with an eye to evaluating the third, fourth, and fifth rationalizations. In particular, we would encourage cross-lagged correlation analysis involving a broader sample of nations and different, probably longer, time periods over which the causal processes are more accurately mapped. A broader and larger sample would permit separate analyses for homogeneous and heterogeneous nations to test Deutsch's hypothesis, a task we could not perform with only seventeen or eighteen nations to subdivide.

DISCUSSION AND CONCLUSIONS

Political institutionalization of party systems in Black Africa seems, in some real part, to be a function of the end of colonial rule. In addition it has some relationship to the occurrence of severe civil strife. The plausibility of a causal relationship with social mobilization is, however, seriously reduced. For none of the three independent variables is the relationship as clear-cut as that hypothesized in the scholarly literature, which still relies predominantly on anecdotal "evidence." To the extent that patterns exist in the directionality of relationship, it can be concluded that the post-independence period has been characterized by greater uninstitutional party activity, but that the direction of process — "decay" or institutionalization — is less predictable on the basis of knowledge of political independence alone. Civil strife, which is hypothesized to affect the post-independence process, itself exhibits mixed results. Highly institutionalized and institutionalizing systems seem to "decay" with strife, while less institutionalized and "decaying" systems begin a process of organization with strife. Together these results might suggest that to the extent that "independence" and civil strife are extraordinary, shocking, or disruptive events, a process of political institutionalization in some contexts is fostered more by environmental stimulation than by environmental stability. The implications of such an inference are obvious and exciting. For one, viable political institutions might be expected to develop more rapidly under conditions of stress. For another, and more inferential, civil strife and conditions of social conflict may contribute to the development of institutions capable of ameliorating conflict, or at least to the institutionalization of mechanisms linking publics and government.

But caution is in order. The selectivity of sampling is a *real* threat to "external" validity and generalizability. Our results may well apply only to particular countries in Black Africa. Another reason for caution is our inability to extricate the direction of causation from the apparently significant relationship between severe civil strife and political institutionalization. While either possibility is theoretically interesting and important, a resolution must wait for more rigorous designs, comparative or quasi-experimental. We believe, however, that we have established the existence of some real relationships while controlling for *many* of the plausible rival explanations of those relationships. Similarly, our results have severely reduced the plausibility of one set of relationships, that involving social mobilization and political institutionalization in new nations.

4

THE DEVELOPMENT
OF SYSTEM LINKAGES
IN THE EUROPEAN COMMUNITY

JAMES A. CAPORASO

In this chapter I would like to assess the extent to which the European Community (EC), formerly the European Economic Community (EEC), has developed internal linkages, ties, and interdependencies. Despite the fact that we have observed an impressive growth on the part of many of the Community's structures, it is entirely possible that linkages between these structures are very weak. Some have said exactly this. Altiero Spinelli (1966) has argued that despite its impressive growth in some respects, the European Community is in fact a loose conglomerate of many different power centers, which hang together in a decentralized and poorly coordinated way. If this is true, it may not be appropriate to speak of the Community as a system at all, but rather as a set of separate systems which are "bonded" only by their being called into existence by the Rome Treaty. Since it is a specified set of linkages—linkages among the component parts of the Community—that will be examined, this chapter cannot be taken as a definitive statement on the existence of linkages in general.

Before moving on to test the extent of these linkages, I want to clarify the two dominant ways in which scholars have approached the concept of integration. On the one hand, integration has been used to refer to the process whereby structures and functions emerge and develop at a new level which is more comprehensive

This is a revised version of an article by the same title which will appear in *The Structure and Function of European Integration,* published by Goodyear Publishing Company, Inc. Reprinted by permission of the publisher.

than the previous levels. Here the dominant focus seems to be that of growth, without much attention to the ways in which these growing parts interrelate. This conceptual orientation seems to be illustrated by Deutsch, Burrell, et al.'s definition of integration (1966), though not by Deutsch's much broader theoretical orientation, by Inglehart's focus on the growth of favorable attitudes toward European institutions (1967), by Bliss's focus on the political development of European institutions (1970), and by Lindberg and Scheingold's emphasis on the growth of a decision-making capacity for European institutions (1970). There is some evidence to suggest that in both biological and social systems, performance capabilities tend to emerge first and that these are later followed by coordination facilities.[1] Although these two dimensions are by no means independent (e.g., a total lack of coordination may not allow the system to exist at all), it is my working hypothesis that emerging systems dedicate a greater portion of their energies to development in their early phases and shift their attention to coordination problems later. It may be that this is indeed a good strategy and that an emphasis on developing the Community's coordination and cohesion at an early date will limit its ability to grow and branch out later.

Integration has also been used to refer to the processes of social coordination. As Brickman has pointed out, those theorists who have tended to work within the systems tradition have generally viewed problems of coordination as important. He notes that since "the systems perspective places stress on the internal relationships of the system, the problem of coordination of subsystems is patently a critical one" (Brickman 1969:2). Talcott Parsons (1951) adopts this perspective when he is led to ask how complex social systems, whose component parts are highly specialized, can avoid operating in ways that do not involve conflict among the systems' parts. Parsons sees part of the problem's solution in the development of a specialized integrative subsystem, responsible for the scheduling and coordination of the component parts. The attention which Haas and others (Haas 1964; Schmitter 1971; Puchala 1970) have directed toward the concept of spillover also reflects an interest in particular kinds of intrasystem linkages, as does Morton Kaplan's interest in integrative processes as "regulatory processes which join systems or organizations with separate

---

1. To some extent this is, of course, necessary. One could illustrate this simply by asking how it is possible to coordinate components which are not yet in existence. The focus is therefore not on which process comes into existence first but on the pattern of asymmetries that exists once the original structures come into existence.

institutions and goals within a common framework providing for the common pursuit of at least some goals and the common implementation of at least some policies" (1957:98). Of course, the specific mechanisms through which coordination is brought about are varied. Socialization processes are responsible for altering the attitudes and loyalties of members of the system in such a way as to bring them into line with values and norms of the collectivity and, hopefully, to dovetail individual motivations with the stated goals of the political system. Allocative processes generally distribute rewards in such a manner as to satisfy significant portions of the public and to acquire the legitimacy of as much of the public as possible. The distinctive aspect of these allocative processes is that they are backed up by a legitimate system of coercive sanctions. Exchange processes attempt to bind together diverse interests by establishing some exchange of valued things between them. Both allocative and exchange processes are distributive. They are distinct in that the former is supported by a system of sanctions. As a result, allocative processes tend to deal with public goods (i.e., those which are in varying degrees indivisible for a given population), while exchange processes deal with private goods (Olson and Zeckhauser 1970).

SYSTEMS AND LINKAGES

For the purposes of this chapter it will be sufficient to view a system as a set of elements (components) which to some extent interact with one another. Interactions are only one of the ways in which a system may be bonded. Others include homogeneity of parts (e.g., morphological similarity), common loyalties of component parts to a third unit, the envelopment of several components by an additional structure, and the sharing of subparts by several units. We will not explore the nature of all of these linkages here but will confine our attention to only one type of bond, namely, interactions. Whatever we may say about the importance of the other linkages, it seems clear that interactions are crucial. Campbell (1958) sees the "common fate" of the system's parts as a crucial element in any empirical assessment of "systemness." If parts of a system are capable of thriving or dying without affecting other parts, we would not be inclined to treat these parts as being very cohesively related to one another. Similarly, Singer defines a system in terms of interdependence, as does James G. Miller.

By a social system, then, I mean nothing more than an aggregation of human beings (plus their physical milieu) who are sufficiently interdependent to share a common fate (Campbell) or to have actions of some of them usually affecting the lives of many of them. (Singer 1971:9)

The meanings of "system" are often confused. The most general, however, is: A system is a set of interacting units with relationships among them. (Miller 1971:271)

Other views of system retain the focus on interdependence but shift our attention to a more cybernetic perspective, as in Buckley's definition of a system as a collection of "elements in mutual interrelations, which may be in a state of 'equilibrium,' such that any moderate changes in the elements or their interrelations away from the equilibrium position are counter-balanced by changes tending to restore it" (1967:9). The crucial point here is that the interactions which characterize the system are not strictly determined. There is some "system slack" (Teune 1971) or, in Miller's terms, a "zone of stability." Miller defines this zone as the "range within which the rate of corrections of deviations is minimal or zero, and beyond which correction occurs (1971:292). Up to a point, changes in the behavior of one part of the system may not call forth responses from other parts of the system. However, once a certain threshold is passed, the operation of another variable is begun. If the operation of this variable works to restore the values of the other variables to their original position, the system is characterized by negative feedback mechanisms and static equilibrium. If a new equilibrium is established, we say that the system is characterized by negative feedback and dynamic equilibrium. On the other hand, if the variable called into play operates to throw the other variables still further away from their goal state, we say that the system is characterized by positive feedback and disequilibrium.

THE LOGIC OF EXPERIMENTATION

In this chapter I will attempt to test the simplest of a class of linkages — those based on linear, single-bonded effects, the products of visible, short-range impacts. There will be no focus on curvilinear effects, more complicated delayed effects, compensatory changes, and multiple-bonded relationships.[2] Since these causes and effects are generally short-term, and since their interpretation is surrounded by many equivocalities, it is necessary that some ground rules be adopted. The ideal design for strong inference would be a "true" laboratory design where one has access to precise laboratory equipment and can randomly assign subjects to treatment and control groups. However, I am interested in "real" social situations where self-selection into groups occurs and where confident controls over potentially confounding variables are usually not possible. Similarly, when one is interested in a social process in which he has little or no control over either the independent variables or the context in which it occurs, there is little hope that unequivocal effects can be demonstrated through invoking verbal arguments. In a richly textured world where all variables are allowed to fluctuate simultaneously, there is little constraint on the conclusions which one can draw. In short, there is little place in research concerning ongoing social and political processes for the equipment and technology of the laboratory, its neat separation of subjects into experimental and control groups, its abrupt, incisive administration of the stimulus (independently of other similar administrations), and its precise recording of the response. We cannot voluntarily manipulate the independent variable, and there is usually a substantial lapse of time between the occurrence of the event in which we are interested and its presumed effect. Finally, since we cannot achieve the closure (isolation) surrounding experimental tests, we are denied the experimenter's luxury of adopting strong assumptions concerning the disturbing influence of potentially confounding variables.

In the introductory chapters, several quasi-experimental approaches were outlined which enable one to approximate the laboratory experiment. In the pages that follow I will adopt one particular type of quasi-experiment, the interrupted time-series design (Campbell and Stanley 1963), and apply it to an assessment of developments within the EC. I will examine three important events: the formation of the EEC (1958), the adoption of a series of agricultural decisions (December, 1964), and the crisis which occurred in 1965.

2. The following is an example of a multiple-bonded relationship. Suppose A and B are two variables that are unrelated. However, A and C are related and B and C are related. This is a case where the system itself may be characterized as interdependent, irrespective of the lack of relationships among some of the component parts. Another way to express this is to say that this represents a higher-order relationship, where the relationships themselves are related.

QUASI-EXPERIMENTAL ANALYSIS

QUASI-EXPERIMENT NO. 1: FORMATION OF THE EEC

On January 1, 1958, the EEC, composed of Belgium, Luxembourg, France, the Federal Republic of Germany, the Netherlands, and Italy, came into effect. The Treaty of Rome set forth the goals of the EEC. These goals included not only the removal of internal tariffs and quotas and the establishment of a common external tariff, but also the pursuit of common policies in a variety of areas, including agriculture, social affairs, and monetary affairs. The founding members of the EEC were interested not only in technical and economic cooperation but also in eventual political integration. Presumably the coming into effect of the Rome Treaty was to set in motion the forces which ultimately were to lead to economic and political integration.

One of the ways of viewing integration is in terms of the expansiveness of the centers around which activity takes place and around which attitudes become organized. Thus one could ask to what degree key national interest groups have shifted their purposes, programs, targets, and tactics to a new, more comprehensive level. To what extent have national interest groups begun to make demands of regional political structures, and to what extent have they come to see their fate as bound up with these structures? From this perspective, the formation of the EEC presents an interesting question: Did the establishment of the EEC act as a stimulus encouraging a shift in interest group activity to the European level? We can begin to answer this question by examining the pattern of interest group formation from 1950 to 1970. Figure 4.1 shows a plot of the frequency of interest group formation from 1950 to 1970.

We can see from figure 4.1 that there is a marked rise of interest group activity at about the time of the formation of the EEC. This activity sustains itself for fourteen time points (three and one-half years) past the formation itself. This would seem to indicate not only that the formation of the EEC had an effect, but that the effect was more than a momentary one. Before moving on to our tests of significance, it is important to note one other salient feature of this plot. The interesting thing, besides the sharp increase in the formation activity after 1958, is the fact that the last two quarters of 1957 show a sharp upswing. This is a phenomenon which may be a general property of social systems. I am speaking of the tendency, in cybernetic systems, for responses to occur before the official occurrence of the stimulus itself. That is, people respond in anticipation of an event.

FIGURE 4.1   Effect of Formation of EEC on Interest Group Formation in Western
Europe

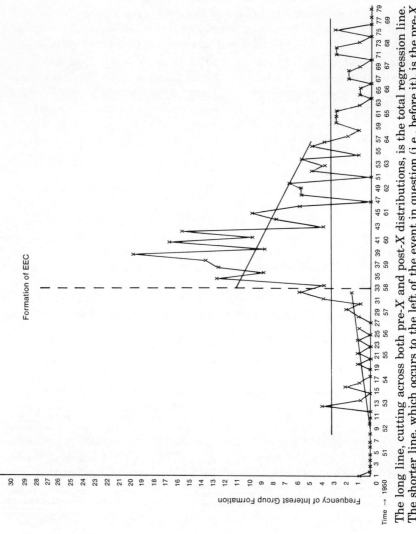

The long line, cutting across both pre-X and post-X distributions, is the total regression line.
The shorter line, which occurs to the left of the event in question (i.e., before it), is the pre-X
regression line. The line to the right of the event (i.e., after it) is the post-X regression line.

Now let us see if our visual interpretation is correct. Table 4.1 shows the results of our four tests of significance.

TABLE 4.1
QUASI-EXPERIMENT NO. 1: FORMATION OF EEC

Variable Name	Pre-* N	Post- N	*T*-Values		Significance Level†			Lag 1 Autocorrelated Error §
			Single-Mood	Double-Mood	Single-Mood	Double-Mood	Adjusted Double ‡	
Interest group formation	32	52	1.47	7.46	yes (.10)	yes (.001)	yes (.01)	.11

* Pre-N and Post-N refer to the number of observations occurring before and after the event, respectively.

† Significance levels are reported for one-tailed tests for values lying between .001 and .10. Values above .10 are not taken to be statistically significant.

‡ This refers to the interpretation of the double-Mood test corrected for the amount of autocorrelated error in the distribution. The tables on which these adjustments were made are provided in Joyce Sween and Donald T. Campbell, "The Interrupted Time Series Quasi-Experiment: Three Tests of Significance," mimeographed, Northwestern University (August, 1965).

§ Autocorrelated error is reported here for the two-segment, separate pretest and posttest, detrended series. That is, we are interested here in the autocorrelation of observations after the regression lines for the pre-N and post-N groups have been separately removed.

The single-Mood test yields a value of 1.47, which is significant at the .10 level. This indicates that the frequency of interest group formation in the first quarter of 1958 could not have been expected on the basis of random fluctuations in the series. The double-Mood value is much larger (7.46) and is significant at the .001 level. This gives us confidence that the establishment of the EEC actually had an impact on interest group activity and substantiates the conventional view of a number of European observers. Lindberg points out that by late 1961, there were 222 economic interest groups organized at the European level, of which all but a few had been created since 1958 (1971:13). Similarly, the *Bulletin der Europäischen Wirtschaftsgemeinschaft* noted as early as 1961:

> The emergence of a European consciousness in the branches of the economy and in the growing interdependence of the markets of the six countries is especially obvious in the great number of occupational associations in industry, agriculture, trade, handicrafts and the free professions which have been established since the ratification of the Treaty of Rome. (1961:11)

Now let us see if we can move beyond the test of significance and assess how well these results stand up to rival interpretations. One possibility is that a general trend in the series is responsible for the increase in interest group activity after formation. This interpretation lacks plausibility because of the general absence of group activity before 1958, with the exception of the last two months of 1957. Since negotiations concerning the establishment of the EEC were going on since 1955, it is not surprising that many interest groups were adapted to the new situation even before it materialized formally. In addition, the hypothesis of trend is neutralized since, even when trend is controlled for in the interpretation of the double-Mood t-value, the results remain significant. Other rival hypotheses are more difficult to cope with. History, or the occurrence of events other than the establishment of the EEC, does not seem a possible alternative explanation, since it is difficult to suggest other important events occurring at that time which could reasonably be suspected of having such an effect. Scoring procedures also present no problem since the data were collected from one standard reference,[3] and since there were no visible changes in recording procedures in the period under consideration. Finally, the rival hypothesis that interest group activity is a poor indicator of political integration is difficult to systematically treat here, since we do not have additional indicators occurring long enough before 1958 to submit them to test. Thus the kind of multiple operationism which can increase our confidence in an indicator will have to wait until later.

We should also note that the EEC formation, in addition to having an impact on the level of subsequent interest group activity, also seems to have introduced a change in the growth rate in that activity. In this regard we are particularly interested in whether there is a significant difference between the slope of the regression line before and after the establishment of the EEC. Walker-Lev test number 1, which tests the null hypothesis that a common slope is suited to pre-N and post-N distributions, is extremely significant, yielding an F-value of 32.67 with 80 degrees of freedom. However, the slope change is in a negative direction, indicating that there was a higher rate of group formation before the formation of the EEC than after. Yet we should be reluctant to interpret this as a negative impact of the EEC. Rather, what

3. The data were compiled from *Répertoire des Organismes Communs dans le cadre des Communautés Européennes* (Services des Publications des Communautés Européennes, 1969).

occurred here probably is that the formation of the EEC had a strong positive impact on the growth of interest groups but that this growth was not sustained over the twenty-year period. In other words, the relationship was curvilinear. A strong, immediate impact resulted in a rapid rate of growth for several time points, then a leveling off, and finally a decline. This pattern is understandable in light of the substantive considerations. One wouldn't expect a constant rate of increase in group formation since there are only a limited number of members and a limited number of issues or areas of concern. Thus, "ceiling effects" come into play at a fairly early stage.

In addition to the slope difference, Walker-Lev test number 3, which tests the null hypothesis that a common regression line fits both distributions, yields a value of 72.27 with 81 degrees of freedom. This value is significant at the .01 level. This primarily reflects the intercept differences between pre- and post-X distributions. This is convincing evidence that both the level at which interest group formation occurred and its rate of growth are significantly different after the establishment of the EEC.

QUASI-EXPERIMENT NO. 2: THE THIRD AGRICULTURAL PACKAGE DEAL

The adoption of the commission proposals by the Council of Ministers on December 15, 1964, was the third installment in the formulation and implementation of the common agricultural policy. The regulations adopted climaxed the end of a long marathon session in which the question occupying center stage was that of a common price level for Community cereals. After tendering several unsuccessful proposals, the commission submitted a package to the council at 5 A.M. Sunday (*Agence Europe* 1964:1). The basic points of agreement in this proposal were the following: (1) an agreement on the Community price for non-durum wheat (425 Deutsche marks per metric ton); (2) West Germany agreed to the ceiling suggested by the commission for compensation as well as to the principle of regressive graduations in compensation; and (3) all agreed to make July 1, 1967, the date for the entry into force of the common price. The importance of this decision for Germany was outlined in a speech by Ludwig Erhard to the Bundestag on December 2, 1964:

> The grain price issue has become the key to further progress in European integration. . . . I know that this step poses grave problems for German agriculture. We take this road with the intention of achieving a breakthrough for Europe, and of making a further

important contribution towards the consolidation of Franco-German
relations. (Erhard 1964:13).

But the importance of these decisions, known as the third agricultural package deal, was not limited to Germany. From July 1,
1967, the date when the common price was to come into effect, the
matter of price policy was to be out of the hands of the separate
national administrations. We will now examine whether these
decisions had any effect on a number of variables representing
economic and political integration. First, let us examine the plot
of one of the dependent variables as shown in figure 4.2.

The visual interpretation of this plot reveals several things.
First of all, the series is dominated by a steady upward growth
trend. There are, of course, fluctuations around the trend line but,
for the most part, these do not appear significant. However, the
slope of this line is neither as steep nor as smooth as the line for
exports. For both of the political measures (decision-making and
adjudication) there appears to be a small drop in the level of
activity after December 15, 1964. Whether this drop is significant
or not is the next question.

We now turn to the tests of significance. Table 4.2 presents these
results.

None of the variables was significant when the single-Mood test
was employed. However, all of the variables with the exception
of the adjudication component were significant on the double-Mood
test. All of these values held up when they were corrected for
autocorrelation.

The interpretation of the F-values presents a different picture:
strong evidence against the hypothesis that adoption of the agriculture package had a strong, positive effect. The slope change
after December, 1964, was significant in three of the five cases,
but in the wrong direction. Of the three cases there was actually a
decline for imports and adjudication and a significant decline in
their subsequent rate of growth. Of course there are strong rival
explanations as to why this occurred. One possible explanation is
that the institutional system of the Community had reached a kind
of saturation point and found it difficult to keep up the old rate of
growth. Countervailing pressures and dampening effects normally
come into operation sooner or later to act as checks on almost all
growth trends. Boulding (1953) has pointed out that many social
and economic variables are of the ogive form, displaying little or
no growth in the early stages, then showing strong spurts and sustained periods of sharp growth, and finally leveling off and perhaps

FIGURE 4.2 Effect of the Third Agricultural Package Deal on the Decision-Making Capacity of European Institutions

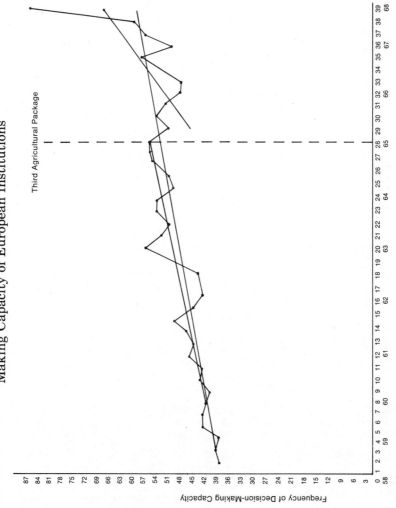

TABLE 4.2
QUASI-EXPERIMENT NO. 2: THIRD AGRICULTURAL PACKAGE

Variable Name	Pre-N	Post-N	T-Values Single-Mood	T-Values Double-Mood	Significance Level Single-Mood	Significance Level Double-Mood	Significance Level Adjusted Double	F-Values Walker-Lev Test 1	F-Values Walker-Lev Test 3	Significance Test 1	Significance Test 3	Lag 1 Auto-correlated Error
T-scored imports	28	20	1.24	4.08	no	.001	.001	48.63	19.05	.05*	.05*	.40
Implementation	28	11	1.26	2.53	no	.01	.01	.15	12.30	no	.05	.19
Decision-making	28	11	1.66	3.02	no	.01	.01	8.77	1.97	.05	no	.33
Adjudication	28	11	1.00	.13	no	no	no	13.34	5.05	.05	.05	.73
Interest group formation	60	24	1.02	3.04	no	.005	.01	3.25	25.54	no	.05	.60

*The tables on which these F-values were checked for significance are calculated to yield critical values at the .05 level. These values are large enough to be significant at a much higher level of confidence.

even declining. This is a good characterization of many of our variables, but particularly the political ones. Another possible interpretation, however, is that it wasn't December, 1964, but June 30, 1965, which precipitated the decline. June 30, of course, marked the beginning of the agricultural crisis in the EEC, and it is plausible that the effects of December, 1964, and June, 1965, are so commingled as to make an independent assessment of their effects impossible. In any case, I think we can conclude with some confidence that the adoption of these decisions did not produce a sustained positive effect.

QUASI-EXPERIMENT NO. 3: ONSET OF THE AGRICULTURAL CRISIS OF JUNE, 1965

The agricultural crisis of June, 1965, which lasted for over six months, was the most severe test for the European Community up until that time. The immediate cause of the crisis concerned some rather technical details related to the financing of the common agricultural policy. However, the underlying causes of the crisis had to do with the proposals of the European Commission that the European Parliament be given increased powers and that the commission be supplied with an independent source of revenue so as not to be tied to the individual states. Additional factors such as the impending majority voting, which was to come into effect by the first of the year, and the emerging role of the commission as a diplomatic actor, may also have played an important part (Newhouse 1967).

The crisis involved the walkout of Couve de Murville, then the French representative in the Council of Ministers, as well as other French representatives of Community institutions where member states are directly represented. These French representatives did not return until the end of January, 1966, when, after two sets of meetings, an "agreement to disagree" was adopted. During the intervening time the five remaining states attempted to conduct routine business as best they could. Most observers agree that the crisis was both a symptom and a cause of a decline in European integration. This assumes, of course, that there was a measurable decline in the first place. Our task here will be to assess how deeply the crisis affected Community behavior on a number of variables.

The plot examining the effect of the crisis on imports and exports among Community members is dominated by a positive trend component. The plot seems fairly smooth and is characterized by a strong upward movement. The autocorrelation coefficients, which are measures of serial dependence in the data, are .99 and .98 for imports and exports, respectively. At about the time of the onset of the crisis, there is no noticeable visual effect for imports; however, in the case of exports among Community members, it does appear as if something has occurred. There is a drop from 56.24 to 54.44 (these data are measures in t-scores ranging from 30 to 80), then an increase to 56 and 58. Then there comes a more sustained decrease, after which it takes twenty-seven months to get back up to the 58.00 level.

As this interpretation suggests, the single-Mood t-value for imports is only .88, which is not significant. The t-value for exports, however, is 3.16, which is significant at the .005 level. Both double-Mood values (2.24 for imports and 5.24 for exports) were significant at the .025 and .001 levels respectively.

The slopes for imports and exports also provide some interesting information. The slope of the import distribution actually increases from .750 to .802 for the post-X observations. Although this is counter to the direction we would hypothesize, it is not enough of a change to be significant (the Walker-Lev value for test 1 is .57). On the other hand, there is a substantial decrease in the slope for the export data, from .996 for the pre-X observations to .531 for the post-X observations. It is also clear from the post-X part of the plot that the slowest part of this growth occurs in the first nine to ten time points (two and a quarter to two and a half years). After that point the growth rate appears to pick up considerably. The F-value for test 1 is 24.61, which is well above the significance level. Similarly, the F-values for Walker-Lev test 3 are 4.53 and 40.56, both significant at the .05 level.

Also, these results are not substantially affected by the presence of trend in the data. The autocorrelation of .18 does reduce the significance of the double-Mood for imports from the .025 to the .05 level, while the autocorrelation of .51 for exports reduces the confidence level from .001 to .01. Although the size of these reductions is substantial, the important point is that the results still lie well within the zone of significance.

Now let us turn to an assessment of the crisis on our three political variables. The plots shown in figures 4.3, 4.4, and 4.5 again give us some visual feel that there is something operating on the system in the region of June 30, 1965. All three variables drop off in the time periods after June 30. The implementation variable dropped from 52 to 50, 51, 51, 50; the decision-making component from 54 to 52, 48, 48, and 52, and the adjudication variable from 61 to 56, 52, 56, and 52. These changes, while perhaps not dra-

FIGURE 4.3 Effect of the Agricultural Crisis on Implementive Integration

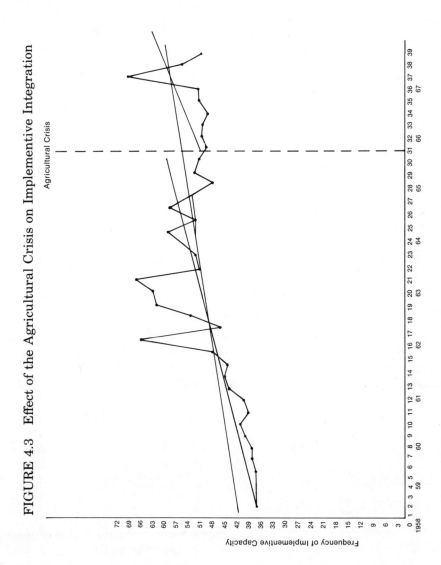

Agricultural Crisis

Frequency of Implementive Capacity

FIGURE 4.4   Effect of the Agricultural Crisis on Decision-Making Integration

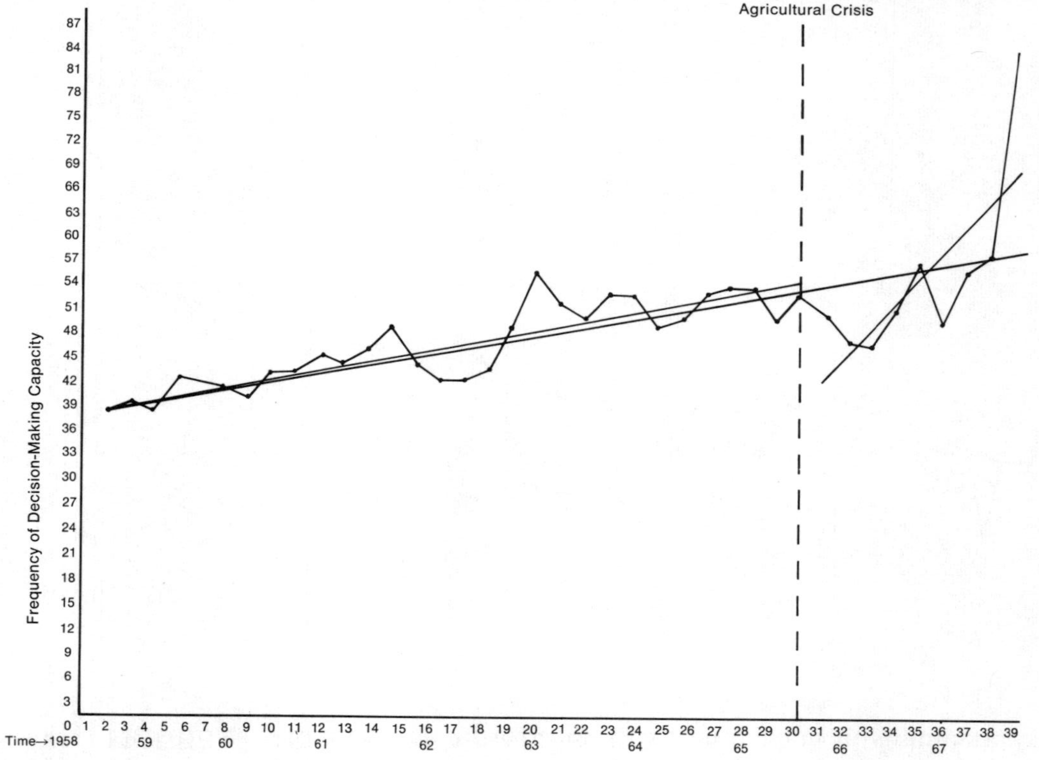

FIGURE 4.5 Effect of the Agricultural Crisis on Adjudicative Integration

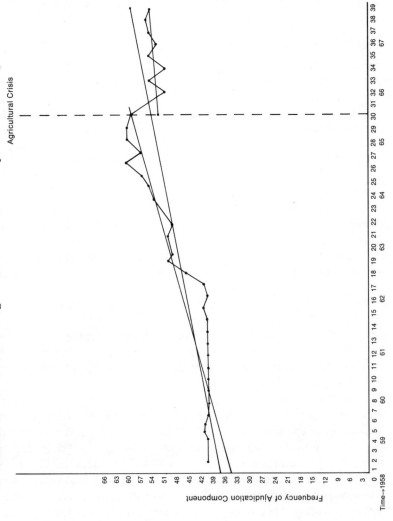

matic, are substantial and, what is more important, they are consistent. This lends strong support to our hypothesized interpretation by rendering implausible the rival explanation of indicator eccentricity. Here we see convincing evidence of convergence in the results of three indicators of political integration. Still, however, we are talking about our visual impressions. What are the results of the tests of significance?

The results from the single-Mood test are again extremely consistent: 1.55, 1.39, and 1.31 for implementation, decision-making, and adjudication, respectively, as shown in table 4.3. The first two of these are significant at the .10 level, while the adjudication values are not quite large enough to qualify. In light of the fact that the largest dropoff occurs in the adjudication variable, this may seem surprising. However, we should bear in mind that it is not departure from the last observed pre-X value in which we are interested but the departure of the first post-X value from the extrapolated pre-X regression line.

The results from the double-Mood test present much the same interpretation. These values are 2.27, 3.69, and 2.77 for implementation, decision-making, and adjudication, respectively. These values are all significant at the stated levels. Again, if we consider how widely these double-Mood values can fluctuate, it is remarkable how similar to one another the results are.

We now shift our attention toward the slopes of the lines after the crisis. The key question here is whether or not the crisis had any effect in slowing down the rate of growth of a variety of Community variables. Let us consider export integration first. The F-value for Walker-Lev test 1 is 24.61, indicating a marked shift in the slope for the postcrisis observations. The slope of the line after the crisis is .531, compared with .996 before the crisis. It is important to interpret these results in conjunction with a visual inspection of the plot. We see that there is a very low growth rate for the first nine observation points following the crisis; for approximately two and a quarter years after the crisis the Community showed no appreciable increase in the rate of export integration. If one considers that the first seven and a half years reflect a period of nearly undisturbed growth, this flattening out of export integration appears all the more remarkable. The fact that the slope after the crisis is as high as it is reflects the extremely sharp increase in export trade in the last quarter of 1967. This increase is sustained in the remainder of our data, right up to the end of 1969. One interpretation here is that by this time the crisis had "worn off" and the Community became susceptible to new influences.

TABLE 4.3
Quasi-Experiment No. 3: The Agricultural Crisis

Variable Name	Pre- N	Post- N	T-Value		Significance Level			F-Values		Significance		Lag I Auto- correlated Error
			Single- Mood	Double- Mood	Single- Mood	Double- Mood	Adjusted Double*	Walker-Lev Test 1	Walker-Lev Test 3	Test 1	Test 3 †	
T-scored imports	30	18	.88	2.24	no	.025	.05	.57	4.53	no	.05	.18
T-scored exports	30	18	3.16	5.24	.005	.001	.01	24.61	40.56	.05	.05	.51
Implementation	30	9	1.55	2.27	.10	.025	.05	.08	8.35	no	.05	.31
Decision-making	30	9	1.39	3.69	.10	.001	.01	17.78	.83	.05	no	.08
Adjudication	30	9	1.31	2.77	no	.005	.05	1.84	25.21	no	.05	.72

* The tables for the adjusted double-Mood are based on critical values for .01 and .05.

† F-values assessed for significance on tables where only the .05 level of significance is given. See Hubert M. Blalock, *Social Statistics* (New York: McGraw-Hill, 1960).

The merging of the Community institutions on July 1, 1967, may have been a factor in this increased integration; the impending completion of the customs union in July, 1968, may similarly have had a triggering effect.

Confidence in the effect of the crisis on economic integration would be increased if the crisis brought about significant changes in other economic indicators. Unfortunately there is no significant slope difference for the postcrisis observations on the import indicator. In fact, though the F-value for test 1 is a nonsignificant .57, the slope after the crisis is actually higher (.802) than before the crisis (.750).[4] However, Walker-Lev test 3, the test of the null hypothesis that one regression line fits both pre- and postchange data, is significant. We can see from the two regression lines that although the slopes are nearly identical, the intercept of the postchange distribution is lower than that for the prechange distribution.

The results for the Walker-Lev tests for our three political integration variables (implementation, decision-making, and adjudication) present an even more difficult problem for interpretation. If we examine table 4.3 we note that there is a significant slope change only for the decision-making variable and that the sign of this change is not in the predicted direction. That is, there is an increase in the slope from .548 for the pre-X group to 3.083 for the post-X group. Again, we must stress that it is crucial to interpret these tests in light of the plot of the variable itself. We notice that there is a decline and then a lull in the decision-making capacity of the Community for four time points (one year) after June, 1965. The values drop from 54 for the last value prior to the crisis to 52, 48, 48, and 52. After this lull the Community seems to become reinvigorated and surpasses its former levels of activity. The point I am trying to make is this: the increased slope after the

---

4. Of course, one way out of this difficulty is simply to adopt an extreme operationalist posture and say that "export integration" is one concept and "import integration" is another. While one is free to adopt this position, it seems to me to have two defects. By making a concept coterminous with its indicator, a great deal of the concept's theoretical power is drained away. Since in this context indicators don't really "indicate" (they define), it becomes impossible to assess the error involved; it thus becomes impossible to assess the discrepancy between a pointer and an underlying concept. Also, if all there is to concepts are their indicators, it appears nonsensical to attempt to deal with these problems in a comparative framework. Indicators are often peculiar to given social settings. The view here is that indicators are partial and fallible attempts to "catch" a concept. It is possible to assess how well this is done by referring to standards of reliability and validity — that is, by the language of measurement.

crisis is to a large extent a function of the low values from which the postchange distribution starts out. The gradual attenuation of the effects of the crisis and the consequent "normal" resumption of the system's growth pattern make the slope that much steeper. The point could be put more concisely and at the same time more generally: if the intercept is not the same or higher for the post-change regression line, it is difficult and probably fallacious to interpret a higher post-X slope as evidence of a true, positive effect of some event.

The slope tests for the implementation and adjudication variables, while not statistically significant, are nevertheless in need of further interpretation. Again, though there is very little difference between pre-X and post-X slopes (.746 and .983, respectively), it should be observed that for a year and a half after the crisis occurred the Community in its administrative capacity operated at a very low level. To some extent this trend may have been under way before the crisis occurred. In any case the slope value again is as high as it is only because of these initially low values. Finally, the adjudication variable shows a sharp drop after the crisis but again picks up slightly as time goes on. The slope actually does decrease from .830 for the precrisis series to .283 for the postcrisis series. However, this decrease is not significant. It is interesting to see that the F-value for Walker-Lev test 3 is 25.21, a value that is significant well beyond the .05 level. The reason for this high value is evident if we note that the intercept in the postchange group is well below that for the prechange group. A similar interpretation could be made for the significance of test 3 on the implementation variable. The intercept for the postcrisis group is substantially below that of the precrisis series.

We can summarize the results of the F-tests in two ways. First, significant results were obtained in six out of ten cases, and in several cases these results were fairly dramatic. Second, we can say with some confidence that it seems to be much more difficult for an event to force a change in the subsequent slope of a series than to force a change in intercept. In terms of substantive theory, this is equivalent to saying that the occurrence of events in the EC has impacts that are generally short-term and limited to affecting the *level of intensity* at which a variable operates, rather than long-term and influential on the growth rate of these variables. To put the matter still a different way, changes occurring in the EC seem to affect other variables but are then "absorbed" by them. They do not seem to have an effect that is cumulative and self-sustaining. In short, while the level of growth and system activity is affected, its growth rate generally is not. Velocity may be easier to alter than acceleration.

For those cases where the plausible rival hypothesis of random instability has been eliminated (that is, those results which are statistically significant), let us inquire further to see how well they stand up to additional threats to validity. The presence of auto-correlation does not affect any of the t- or F-values to such an extent as to make any of them insignificant. In all cases the p-level for the double-Mood values is decreased: for imports, from .025 to .05; for exports, from .001 to .01; for implementation, from .025 to .05; for decision-making, from .001 to .01; and for adjudication, from .005 to .05. However, the effect of this adjustment for auto-correlation is not to push these values outside of the acceptable significance levels. In addition, there do not seem to be any serious *substantive* explanations that compete with the crisis as an explanation of the changes in the vicinity of 1965–66. One possibility, raised peripherally by Miriam Camps (1966), is that both the crisis and the stagnation of European integration from 1965 to 1968 were the result of the unresolved differences between the French and the other five nations over the form and desirability of political integration. The French advocated a loose, confederal scheme along the lines suggested by the Fouchet negotiations, while the other five had something more "supranational" in mind. The reason that the year 1965 was so significant is that by this time the Community had gone a long way toward completing what has been called "negative integration" (Camps 1966:115; Pinder and Pryce 1969). In other words, the Community had been mostly occupied with removing tariffs and internal barriers to the flow of goods and services. In addition, much of the activity occurring during this period was essentially implementive in nature and involved merely the administration of steps that were already carefully defined by the Rome Treaty. Finally, by 1965, the key decisions had been made about the common agricultural issues remaining, e.g., the financing of the policy, the key aspects of the controversy relating to the uniform price system, and the market organizations had been resolved by late December, 1964. Although a great deal remained to be done in such areas as monetary matters, transport, and social affairs, the Community had run out of steam in those areas in which there was basic agreement at the time the Rome Treaty was signed. Therefore, as Camps observes, by 1965 the EEC had reached a watershed in which it was forced to tackle more controversial issues – issues whose resolution had uncertain political effects. The French were very reluctant to become involved in the formulation of economic policies which might involve political integration.

While this interpretation of the crisis is convincing, it does not really affect the validity of our interpretation that the crisis had an influence on subsequent Community integration. It merely puts these events in a proper developmental sequence, i.e., from political disagreement to crisis to Community disintegration. However, a rigorous test of this is not possible here. Essentially, it would require holding constant or "partialing out" the effects of the crisis. If the postcrisis disintegration (or lull in integration) still obtained, the conclusion might be that these underlying political disagreements caused the crisis and disintegration. The situation would be described by the first rather than the second diagram below, where A = underlying political disagreement, B = the crisis, and C = postcrisis disintegration.

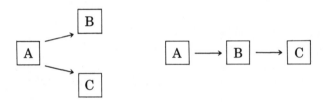

Finally, a word seems necessary about the rival hypothesis of irrelevant indicators. Essentially, we have used Community regulations, decisions, directives, and a variety of additional official Community acts to indicate the level of decision-making in the EC. A convincing argument is offered (e.g., Macrae 1970) that although the EC kept up the pace of its activity, its style after the crisis was considerably more timid and its autonomy considerably more restricted. In the place of the previously bold commission initiatives, there is substituted now a petty bargaining mentality (*un esprit de marchandage*) and a shift in the locus of power to the Committee of Permanent Representatives, the delegate body of the nationally oriented Council of Ministers. If we were to take this argument seriously, and I think it deserves grave consideration, it would add confidence to the interpretation that the crisis did have a serious impact on the level and rate of development of subsequent integration. It would indicate that we are working here with indicators which are not as sensitive as they might be. We would thus be in the position of confirming our hypothesis through use of the "hard case." We could argue that confirmation could be more dramatically made through indicators which got at the autonomy and forcefulness of the commission in the decision-making process.

ASSESSMENT OF THE MODEL AND REFORMULATION

The original model on which this analysis was carried out was based on the assumption of nonindependent observations over a period of time. "Change" in this model was assessed through the criterion of whether a series displayed discontinuous behavior in the region or posterior to the region in which an important event occurred. A "true" change in this model is reflected in some value (or values) of a variable that departs significantly from an actual or projected regression line. It should be emphasized that this model makes no assumptions about the time required for the event to register an effect or for the relative weight of impact of an event on the beginning, end, or any other point of the series. It is assumed that once the event registers its effect, its impact is felt uniformly throughout the series. There is thus no provision in the model for "wearing off" effects or the gradual erosion of an event's impact over time.[5] There is, in addition, no provision for curvilinear effects or other complex, functional relationships. We now want to evaluate the results of this analysis and suggest how their interpretation stimulates a reformulation of this model.

The first conclusion to draw is that, on the basis of this limited analysis, the European Community appears to be a moderately well-integrated system. There is, over a fairly broad, nontrivial range of indicators, a reasonably well-developed set of linkages. These results are seen as partially superseding those presented elsewhere.[6] Let me attempt to summarize these results. Most of the dependent variables we examined reacted to (i.e., were sensitive to) the three events. There were, however, significant variations in these patterns of response. The single-Mood statistic was

5. A "wearing off" effect is one where the causal event actually weakens (and gradually extinguishes) with the passage of time. Since the event is conceptualized as a discrete occurrence, it is assumed that its continued influence is a function of some trace or residue it deposits in the system, and that such traces become increasingly less operant with the passage of time. An "erosion" effect, by contrast, is one which retains its full (absolute) force over time before its causal impact is decreased because of the incorporation of additional variables into the system.

6. The results presented by Caporaso and Pelowski (1971) justified the conclusion that the EEC was only a "weakly integrating system." The differences are admittedly ones of degree and are probably attributable to the presence of more stable, composite indices here. The analysis in the cited article was carried out with single indicators as dependent variables.

significant in four out of eleven cases and the test for common slope (*F*-test number 1) in only five out of ten cases. By contrast, the double-Mood and the test for the null hypothesis of a common regression line (*F*-test number 3) were significant in ten out of eleven and eight out of ten cases, respectively. What this means in substantive terms is that in most cases the occurrence of an important event had no immediate effect and in only half of the cases was there a significant change in the form of the relationship. Yet there is overwhelming evidence to suggest that pre-*X* and post-*X* distributions are significantly different. The double-Mood test and Walker-Lev test number 3, both of which are based on comparisons of regression lines, provide strong, convergent evidence for this. I would like to suggest a reason for this and then attempt to reformulate the original model.

One reason why the test for slope failed to show significant change more often may become clear once we recognize the typical response of a variable to the occurrence of an event. Let us take the case of the effect of the agricultural crisis on decision-making in the European Community. The plot for this is given in figure 4.4. We can see that there is a marked decrease, which lasts for approximately one year, after the crisis. After that time it appears as if the Community resumes its course. Let us consider for a moment the effect of this type of behavior on the interpretation of the slope test. The predicted direction of the slope change is, of course, negative; that is, we expect the rate of growth of decision-making to be less after the crisis. As a matter of fact, the slope is steeper, as can be seen by examining the regression lines drawn in figure 4.4. My interpretation of this is straightforward. The crisis did have a predicted (negative) effect, but it was not a permanent one. If one examines the slope for the first four or five time points after the crisis, one does find a negative slope. The effect of the crisis seems to have lasted for about one year, after which time the Community resumed its normal decision-making growth pattern. (By "normal," we mean normal with respect to an altered baseline.) Thus it is assumed that the crisis was permanently absorbed into the operation of the decision variable in the form of a lower level or frequency of operation, even though the negative growth rate seems to have completely eroded after one year.

AN ALTERNATIVE MODEL

The utilization of a theoretical model has several advantages. By setting forth theoretical expectations, it provides a set of baselines useful in interpreting interesting substantive results. Second, by establishing (before data analysis) a fairly simple and explicit set of functions to which the data should conform, it places some restraint on opportunistic and ad hoc curve-fitting. Third, and most important here, our enunciation of the expectations of the model provides us with clear standards in terms of which to assess un-expected results. It is these unexpected results which are re-sponsible for announcing defects and inadequacies in the original model, thereby making reformulations and improvements possible.

Minimal requirements of the reformulated model would have to include the following: (1) it would not specify any immediate effect posterior to an event, thus allowing for the joint possibilities of delayed or anticipated effects; and (2) it would provide for a temporary effect, typically lasting one or two years and then eroding, possibly by becoming extinguished or by being absorbed by the system.

These kinds of expectations are paralleled very closely (though in a much more formalized fashion) by a type of stochastic model formulated by Box and Jenkins (1962) and Box and Tiao (1965) and labeled the "integrated moving average process." The core of this model is that of a system (or variable) subjected to periodic random shocks, a proportion of which are absorbed into the level of the series (Box and Tiao 1965). An additional property of the model is that the farther one moves away from the event in time, the less the behavior of the distribution is a function of the event.

In terms of substantive theory, this process most closely resem-bles something lying about midway between what Boulding (1970) calls the equilibrium and cumulative processes. This process is one in which a disturbance variable forces an adjustment by the sys-tem, after which equilibrium is restored. However, this equilibrium is at a new position, having incorporated a trace of the disturbance into its permanent behavior.

A final word may be said about integration in Europe, as growth and as coordination. All the variables examined, with the exception of interest group formation, are increasing and have been in-creasing for the past ten years. Thus the European Community could be conceptualized as a positive growth system, displaying impressive increases across a broad range of variables. However, the coordination of these components, as measured by the sensi-tivity of selected system variables, is somewhat less impressive, though by no means negative. Nevertheless, our assessment at this point is that the growth component is the dominant one. This conclusion should not be a cause for disappointment. It may be a good strategy for a "young" system to avoid excessive centralization and coordination, since there is likely to be a tension between growth and coordination. Growth rests on a niche-filling strategy, an ability to probe and respond to the demands of a changing environment. It thus requires a great deal of freedom to stumble, to find, and to exploit the proper opportunities. Seen from this perspective, the looseness or haphazard fashion in which Community structures sometimes interrelate is one of the Community's strongest points.

5

LINKAGES BETWEEN WAR AND DOMESTIC POLITICAL VIOLENCE IN THE UNITED STATES 1890–1923

MICHAEL STOHL

It has become rather commonplace when writing about violence in America to suggest that Americans suffer from what Graham and Gurr (1969:788–822) call "historical amnesia," or a repression of violence in the national consciousness (Hofstadter and Wallace 1970). Although reasons for this amnesia may be quite complex, there is clearly a conceptual blinder which assists this repression. American scholars generally consider the federal government as an instrument which protects the community from violence, although the "greatest and most calculating of killers is the national state, and this is true not only in international wars but in domestic conflict" (Hofstadter and Wallace 1970:6). The sources of this conceptual problem are found in the distinction made between government (with its political structure) and society at large. The use of physical force has been regarded by many writers as a characteristic and hence, by implication, a legitimate feature of the political structure. Following Weber, we may identify the state as

> a relation of men dominating men, a relation supported by means of legitimate (i.e., considered to be legitimate) violence. (Gerth and Mills 1958:78)

A government is thereby identified with legitimate use of force, and it becomes natural to dismiss governmental violence because

156

it is simply viewed as a normal governmental function. Research on violence within societies has generally therefore concentrated on "abnormal violence," or what is commonly referred to as civil strife.

Nardin (1971) has incisively discussed the sources of this conceptual problem. It is worth summarizing his discussion, which concerns the identification of the state as either a conflict manager, as Marx perceived the Hegelian conceptualization, or a party to conflict. In the vulgarized Hegelian conception:

> Conflict becomes the clash of special interests which it is the function of governments to resolve, and violence a form of partisan action which it is the government's task to manage with force, if necessary. (Nardin 1971:14)

Marx, however, contends:

> Government, far from being above the special interests in society, in actuality constitutes a weapon employed by the dominant interests against the weaker. The State always represents the interests of the ruling class. (Nardin 1971:14)

In the latter approach the state's use of force is seen as the violence of a partisan faction, not as conflict management.

I have emphasized this particular problem of the conceptual scheme because it is my belief that it has also hindered the study of connections between domestic and international affairs. Social scientists have created analytically distinct subfields in their pursuit of theory building. One of the most familiar of these distinctions involves the study of political processes within and between states—between comparative and international studies. The area where these two systems overlap has been identified as the study of linkage politics (Rosenau 1969). The study of linkages attempts to identify and assess those areas where two or more analytical systems overlap, and the relationships that are so produced. Thus, linkages are defined as "any recurrent sequence of behavior that originates in one system and is reacted to in another" (Rosenau 1969). A number of earlier studies have also been interested in such linkages (Sorokin 1937; Rummel 1963; Tanter 1964, 1965; Collins 1969; Wilkenfeld 1968, 1969; Stohl 1971). Due to the differences in their conceptual approaches and methodological analyses, these studies have produced contrasting results (Stohl 1971). However, they do share a number of conceptual problems despite their differences.

They have all employed a model of conflict management which maintains a distinction between conflict management within and among nations. The dominant view is of the state as a conflict manager within the territorially bounded nation-state, and merely as a member of the "Hobbesian state of nature" in the international realm. An alternative construct is advocated in this paper. Conflict management is the process of

> reducing the violence of conflict between groups, including conflicts between political authorities and other members of the community or between rival authorities within national societies or in the international system. (Nardin 1971)

Thus we will consider all collective violence that occurs in a society rather than that directed against the "legitimate authorities." Adoption of this conceptualization of the state as party to conflict behavior facilitates the bridging of boundaries between the study of conflict within nations and the study of conflict between nations by proceeding with similar conceptual frameworks in both systems.

A second problem, which may have its sources in the first, is that these studies have dealt with conflict behavior in general without separating violent and nonviolent behaviors. This need was identified by Weber:

> The treatment of conflict involving the use of physical violence as a separate type is justified by the special characteristics of the employment of this means and corresponding consequences of its use. (1947:133)

That is, previous studies have compounded the error of dealing solely with partisan violence against governments by that of considering any collective conflict behavior that occurs, whether or not violence took place. Thus strikes, demonstrations, and protests which have no violence associated with their immediate occurrence are generally pooled with similar occurrences which do result in violence, whatever the cause. It might be argued that it is an empirical question whether this lumping of violent behavior with other conflict behavior distorts the results; nevertheless, the answer cannot be ascertained unless we distinguish between violent and nonviolent conflict behavior and compare the results.

A third problem is the recent tendency for this type of research to be of a synchronic nature (dealing with phenomena that take place with many units but within the same time period) rather than diachronic (dealing with phenomena that take place over many time periods but within a limited space region). While for some subjects of research the approach may not be crucial, attempts to determine the impact of events in one system on those in another should be concerned with interactions and processes of change over time.[1]

The study of the interaction between domestic and international systems must make use of a conceptual framework and a methodological approach which are conducive to the study of linkages. The framework to be discussed in this paper assumes that the political system is the distributor of values within a social system. The approach taken is consistent with a group conflict theory orientation in that its basic premise is that "violent conflict, and revolution, arise out of group conflicts over valued conditions and positions" (Gurr 1971:12). In the section below we will examine the interaction between violence in the domestic and international systems. To do so, it is first necessary to outline a plausible model of the structures and interactions in the domestic system, and to discuss how and with what effects violence in the environment (including the international system) is likely to alter the structures and interactions internal to the system.

THE UNITED STATES AS A STRATIFIED SOCIAL SYSTEM

The development of a political system and corresponding structures which distribute societal values implies a choice as to what is valued by the members of the political system and on what basis values should be distributed. In political systems characterized by individuals and groups having compatible values or value agreement (congruence), politics would simply be the administration of valued things. In political systems characterized by conflicts of interest—conditions in social relationships in which individuals or groups hold incompatible or mutually exclusive values—politics is the distribution of values by authorities. Weber termed the structures developed to perform these functions "political," "in so

1. In earlier research I have found that these synchronic studies have also obscured the relationship between internal and external conflict behavior when economic, political, sociocultural, or technical capabilities of the nations in the sample studies are not controlled. Thus no relationship between internal and external conflict behavior was found when the entire population of nations was analyzed collectively, but when nations were grouped by political structure (Banks and Gregg 1965) and the same information reanalyzed, personalist and polyarchic nations revealed moderate correlations while only centrist nations revealed no relationship (Stohl 1971).

far as the enforcement of its order is carried out continually within a given territorial area by the application and threat of physical force on the part of the administrative staff" (1947:154). Either individuals or groups who do not receive the rewards they want will voluntarily comply with authoritative decisions, or authorities will attempt to force them to comply.

The decision of what to value implies inequalities among values and a system of ranking these values in order of the importance of their distribution. Parsons (1953:97) suggests that stratification is "the ranking of units in a social system in accordance with the standards of the common value system." There may be natural differences of age, sex, or race among individuals, but clearly there is some structure that translates these individual differences into social inequalities. The position taken here is that as a result of a society's stratification system, enforced by the political structures, individuals are accorded places in various categories or groups. Some social structures may thus usefully be seen as segmented. A segmented system has attribute groups whose members are readily identified by a number of established criteria.

> Segmented society, therefore, is the polar opposite of individualistic society – a society characterized by the apparent interchangeability of all inhabitants. (Rogowski and Wasserspring 1972:37)

To illustrate this concept, let us conceive of a very simple stratification system where rank was accorded on the basis of race only, and where there were only three races: black, white, and yellow. Let us further suppose that there were only three strata or ranks in the system (analytic levels of stratification systems composed of one or more segments and/or individuals based on the combination of one or more value criteria) – top, middle, and bottom; and that whites occupy the top, blacks the middle, and yellow the bottom. In this white dominant society a black man is easily identified by his racial attribute and ranked according to the place provided to his segment in the population (middle stratum). In order to change that rank (and hence the distribution of valued things), he would have to change his skin color, a process involving high psychic as well as real costs, or band together within his attribute group in order to collectively attempt to change the ranking system. This could occur in isolation or in conjunction with the yellow segment. If this attempt were successful, the new dominant segment would include the black segment, and the individual would have moved up the ranking system and increased his relative share of valued things.

The nation may thus be seen as a stratified social system. Adopting this approach, four working assumptions concerning behavior in status systems need to be explicated (Rummel 1971b).

1. Behavioral dimensions are linearly dependent on status.
2. Segment behavior is directed toward higher-ranking segments, and the higher a segment's rank, the greater its status behavior.
3. High-rank segments support the current social system.
4. Segments emphasize their dominant rank and other subordinate ranks in interactions.

The first assumption is the key notion of status theory — that status shapes behavior and can be specified by a linear relationship.

> This linearity is intuitively sensible and status theory provides no reason to assume otherwise. Status itself is a linear continuum running from low to high (most desirable) and the higher a status the more a particular behavior. And this linearity appears to be confirmed by empirical results. (Rummel 1971b:50)

If status is linear and desirable, then segments will want to move toward the top of the status hierarchy. The closer they move toward the top, the greater will be their interest in maintaining the system which accords them their status rewards. At the same time, since they have acquired status and the system interaction is based on status, they will emphasize their status positions in interactions with those below them in the status hierarchy. As Coser has said, "to the vested interests, an attack against their position necessarily appears as an attack upon the social order. Those who derive privileges from a given system of allocation of status, wealth and power, will perceive an attack upon these prerogatives as an attack against the system itself" (quoted in Rummel [1971b:53]).

Within this framework it is possible to distinguish six types o conflict behavior in which these value incompatibilities may be manifest:

1. *Reaction:* Attempts to decrease the relative distribution of values to lower segments and increase the distribution of values to the dominant segments.
2. *Conservatism:* Attempts to maintain the status quo in the face of changing conditions.

3. *Accommodation:* Attempts to increase the size of the dominant strata through individual or segmental mobility.
4. *Reform:* Attempts to increase the distribution of values to lower strata and decrease the distribution to dominant strata.
5. *Radicalism:* Attempts to change the rankings of segments and thus the distribution of values.
6. *Revolution:* Attempts to change the value system and the replacement of one value system with another.

We have stated that the ranking system is determined by the dominant segments' ranking of the remaining segments. Does this imply that there is a "power elite," in C. Wright Mills's terms, that conspires to decide what the system will be? Pluralist democratic theorists would have us believe that the major organized interests in society, and most notably capital and labor, compete on equal terms, and that no single group is able to achieve any permanence in positions of power. The groups thus continually compete with each other for the right to rule. E. E. Schattschneider has commented on the pluralist perspective:

> The flaw of the pluralist heaven is that the heavenly chorus sings with a strong upper-class accent. . . . The system is skewed, loaded and unbalanced in favor of a fraction of a minority. (1960:31)

It was this unbalanced, upper-class accent of the American power structure that led Mills to describe the power elite. There has been too much evidence of mobility, and not enough of conspiracy, within the upper strata of society to allow for a strict power elite interpretation. But the fact remains that there has been a continuous domination of a few segments of society over the rest. E. Digby Baltzell was the first to systematically document the composition of the American social upper class and their backgrounds. From his study (Baltzell 1958) it can be seen that the members of the social upper class and, correspondingly, the social register are also rich businessmen and their descendants, although the register is unlikely to contain members of the ethnic rich, such as Irish Catholics and Jews. However, there has been much movement within the middle strata and even entry of some of the upper middle class into the top strata. At the bottom, however, there is a very different phenomenon, "a pariah group that is collectively identified and denied social mobility either individually or as a class. It is an ascribed as well as an achievement criterion that maintains these groups at the bottom" (Jenkins 1969:130).

The important question is whether we can determine if certain identifiable segments are indeed in favored positions in the American social structures. A ruling elite model would imply that the same persons control a wide variety of issues. We posit, instead, a governing strata model which merely implies that there are a number of segments which receive a disproportionate share of the country's income, own a disproportionate share of the country's wealth, and contribute a disproportionate number of its members to the controlling institutions and key decision-making groups in the country (Domhoff 1967). Below we distinguish four major ranking systems and the approximate order of prestige within each rank, and suggest that there is a pattern to the holding of dominant positions in American society.

Racial	*Ethnic*	*Economic*	*Religious*
White	Northern European	Employer	Protestant
Black	Southern European	Landlords	Catholic
Oriental	Eastern European	Employees	Jewish
Indian	Latin	Farmers	
	African		
	Asian		

What this suggests is that white, Anglo-Saxon, Protestant employers are the holders of the positions in the dominant segments in society. With few exceptions this has been the case. The majority of the exceptions have involved landlords and large landholders in the South and West, where manufacturing was not as important as in the East. But in terms of the holders of offices of national power and the control of major U.S. corporations, there has been a predominance of families residing in the northeastern portion of the United States who were white, Protestant, and successful businessmen (Domhoff 1967; Baltzell 1958). For a graphic illustration of this phenomenon in one city, the reader should consult Warner and Lunt's Yankee City studies (1941: 225, table 7).

The government reflects the dominant ranking of the segments in the important subsystems. The government itself may also be seen as a ranking system. At the top of this system is the U.S. federal government. We know that there are often disputes between federal and state or local governments. The case of school integration illustrates one conflict between national and state policy. In the state of Mississippi, federal troops assisted James

Meredith in enrolling at the University of Mississippi against the will of the Mississippi state government.

What this implies, then, is that the dominant segments, through their disproportionate share in the nation's wealth, prestige, and power, in large part determine the continuation of the segment ranking system and hence the resultant distribution of goods and services. This existing distribution of values in large part determines potential future contribution in both positive and negative terms. This determination of values thus provides the major basis for political conflict and violence.

The interactions involved in attempts to maintain or adapt the rank system to changes in the environment involve the systems concept of morphostasis and morphogenesis.

> The former refers to those processes in complex systems environment exchanges that tend to preserve or maintain a system's given form, organization or state. Morphogenesis will refer to those processes which tend to elaborate or change a system's given form, structure or state. Homeostatic processes in organisms, and ritual in sociocultural systems are examples of morphostasis; biological evolution, learning, and societal development are examples of morphogenesis. (Buckley 1967:58–59)

In terms of the conflict patterns elaborated earlier, reaction, conservatism, and accommodation would be examples of morphostasis; reform, radicalism, and revolution would be examples of morphogenesis, since they change the rank system of a society or create a new system.

Changing economic conditions may cause the potential contributions of a segment and the distribution of goods among segments to rise or fall. This would create a rank disequilibrium, in Galtung's terms (1964), which would increase the tension in the system to the point at which the dominant groups would have to adapt the system, increase the available goods, or put lower strata in a position to attempt to alter the current distribution of goods and rankings.

The American economic system, as a capitalist system, has within it the seeds of such problems. It is not only the Marxists who have seen this. The entire thrust of liberal economists has been to create the conditions under which the internal economic system would be maintained within the current systems. The members of the American dominant segments have thus often seen economic expansion as the key to domestic prosperity and social peace. William A. Williams, referring to the American system, has said:

> They were primarily concerned with obtaining markets for surplus manufactured goods and venture capital, and with acquiring access to cheap raw materials needed by the American industrial system. That industrial orientation became increasingly clear during the twentieth century as American leaders struggled to build and maintain an international system that would satisfy the interrelated economic, ideological, and security needs and desires of the United States as they defined those objectives. (1970:4)

As a result of the attempted expansion of the free marketplace, the United States became involved in a number of wars—"wars to apply the principles; and wars to defend the freedom and the prosperity that the expansion the principles had obstensibly produced for Americans" (Williams 1970:46).

In this paper we will investigate the period from 1890 to 1923, a period in which the United States was involved in two wars, to determine if these periods are characterized by increasing violence as lower segments clash with higher segments over the distribution of scarce values. In the periods following war the dominant segments should be strengthened as a result of the increases in markets and their ability to provide more goods within the same ranking system. However, another result of the participation in war should be an increase in social dislocations (internal migration, upward mobility for semiassimilated segments) which both intensify demands and threaten the positions of the dominant segments. The results should be an increase in conflict and violence between top and bottom. As a result of obtaining the war objectives in these two cases, the dominant segments will have increased goods to distribute in addition to their increased status and will attempt to protect the gains of the war against those who seek to have the system adapted to provide a different distribution. Periods after war would then be expected to show a continuation of domestic violence, with increases in the amount of violence between top and bottom, and perhaps with more being directed downward from above.

> Every successful imperialist policy of coercing the outside normally—or at least at first—also strengthens the domestic "prestige" and therewith the power and influence of those classes, status groups and parties under whose leadership the success has been attained. (Gerth and Mills 1958:170)

PLAUSIBLE RIVAL HYPOTHESES

Summarizing the above, the following hypotheses concerning the ways in which war stimulates changes in the internal political system and results in conflict and violence are postulated.

1. Wartime economic mobilization brings new groups into the productive process and enhances the positions of these groups relative to the dominant segments, thus intensifying conflict and violence.

2. Wartime social mobility increases the status positions of underdog social groups, relative to the dominant segments, which increases the hostilities between them.

3. The economic and social changes of the war generate demands for the reallocation of political power and rewards, which intensifies conflict and violence. In addition, success in war increases the power and prestige of dominant groups and intensifies their efforts to maintain control of the political system (see figure 5.1).

Observation of the impact of war on the type and extent of domestic political violence requires a research design that will allow for the introduction of the independent variable only at specified times in the sequence of events. The subject of study does not allow us to introduce the independent variable (war between nations) as we would be able to do in a true laboratory experiment with individuals or small groups, or within the context of a simulation exercise. To test the impact of war, we will employ the interrupted time-series quasi-experimental design described in chapter 1.

The object of this design is to determine whether a nonrandom change has occurred in the dependent variable(s). Explanations of what change has occurred, and why, can only follow after it has been established that a change has indeed taken place. The change that we assess in the significance tests is the difference between the expected and observed value points or distributions in the months before and after the experimental input. The tests used and their descriptions may be found in the introductory chapter in this volume.

As indicated above, one important way of establishing the credibility of hypotheses is to subject them to a series of plausible rival hypotheses which challenge the explanations offered. Next we distinguish some likely trouble spots and delineate how we intend to guard against these threats to the reliability and validity of the analysis.

FIGURE 5.1. A Simple Model of the Impact of War on Domestic
Violence

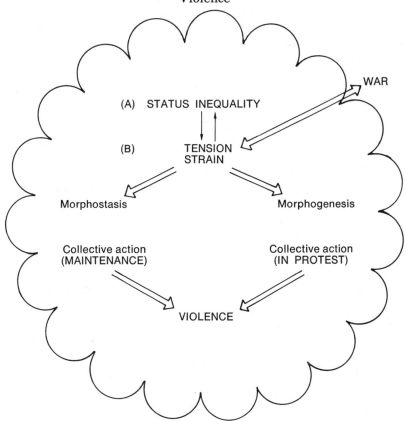

DATA COLLECTION AND CONSTRUCTION OF INDICATORS

There will be two quasi-experiments in this study, each at the time
of U.S. participation in war. In the period of the study (1890–1923),
the United States was involved in the Spanish-American War and
World War I. The dependent variables to be studied are the type
and extent of domestic political violence. *Domestic political violence will here be defined as all collective, nongovernmental or
governmental attacks on persons or property resulting in damage
to them that occur within the boundaries of an autonomous political system.* For operational purposes, "collective" refers to twenty
or more people. This will exclude most criminal acts and juvenile
delinquency, yet it will encompass the majority of political vio-

lence. The phrase "within the boundaries" is meant to exclude such events as raiding across international boundaries and attacks mounted abroad by dissident exiles.

Types of violence are differentiated by dividing violence according to the ranking system within which the violence occurs and according to the focus of the attacking groups. The systems that are considered are the racial, ethnic, religious, economic, and political organizational systems. The events within the first three systems were combined into a social type, and the events within each of the types aggregated by totaling monthly scores. The initiator of attack dimensions is concerned with distinguishing between government-initiated clashes, antigovernment-initiated clashes, and clashes where there is no clearly indicated initiator of attack.[2] The extent or magnitude of violence refers to two dimensions, duration (number of total days that the event lasted) and intensity (the cost of the event in terms of deaths or arrests). Property damage and injuries were to be included in intensity; however, the reporting was very irregular on these two variables

2. There are always problems with any attempt to categorize complex phenomena in a limited number of categories. In addition to the problems typical to any coding scheme, there are a number of conceptual problems that need be mentioned concerning what is, and what is not, "political violence." The following examples of violence will not be examined in this study:

A. Are certain arrests for murder political? If a political figure is convicted of murder (or, for that matter, any major "nonpolitical" crime), is it a criminal or a political act? Is the conviction concerned with political or criminal violence? Two familiar examples of this problem are the conviction and execution of Joe Hill for murder during the labor struggle at the beginning of this century, and the trial of Bobby Seale for murder of a Black Panther in Connecticut.

B. Thomas Szasz suggests that the act of committing individuals to mental institutions is a political crime. In the United States this criticism has been scoffed at by many. However, we do know that the Soviet Union has employed this means to rid itself of some of its most eloquent critics (Mhedyev, for one), and Szasz argues that this political crime occurs in the U.S., though not perhaps at the specific order of the government.

C. How does one classify the arrests of "political" individuals for drug possession? Is this "normal" law enforcement, or a political arrest? Conversely, is it political protest to smoke marijuana if more than twenty people are together — and, if they are caught at one time, is it a political arrest?

D. Studies of criminal law enforcement in the United States suggest that many of the arrests of blacks and other low-ranked individuals occur in large part because of their blackness or low status. Moreover, in certain crimes such as rape, the pattern has been that individuals are more likely to be arrested if the rape is of a white woman rather than a black, especially if the rapist is a black man. Should this be considered an area of political law enforcement, and, if so, does rape sometimes become a political crime, as Eldridge Cleaver suggests?

and they were dropped from the analysis. The data used have
been compiled from the index of the *New York Times*. All events
that appear to fall under the definition of political violence above
have been coded for all available information. This information
is supplemented where possible by information in the stories themselves and by a comparison of the events found in the index with
a number of historical accounts and documentary studies of the
period. While I am aware of the many objections made to the use
of the *New York Times* as a data source, I am also reasonably
confident of the comprehensiveness of its coverage for this particular study.[3] Tilly (1969) suggests in his study of France that the
newspapers provided the fullest enumeration of events and that
their chief bias was toward the overreporting of events in the large
cities. This tendency to overreport events in large cities may be
true of the *Times* coverage as well, but if we are aware of this we
can adjust for it by introducing an appropriate constant. In addition, in a recent study which investigated the accuracy of the *New
York Times* for this type of research, McCormick was unable to
recommend anything of greater value (McCormick 1970).

QUASI-EXPERIMENTAL RESULTS

SPANISH-AMERICAN WAR

The working hypothesis was that each of the "events" would provide a socially given behavior that could be interpreted in the
quasi-experimental terms of the design. The introduction of war
is the event to be assessed. There is nothing else "in" the event
other than the interruption point. The problem of interest is that
of determining the degree of discontinuity around the interruption
point (the war). In April, 1898, the United States declared war
on Spain. If the level of domestic violence was influenced by this
event it is hypothesized that the impact may be assessed by evaluating the differences (if any) between the pre- and posttest slopes
and intercepts. The influence on the type of violence may be assessed by comparing changes in the levels of the different types.
The war (event) was introduced and assessed for impact on the
domestic violence series at the start and end of the war. In addition, the war months were deleted and the pre- and postwar

3. I am, however, not as confident as the editors of the index, who suggest in
their introduction that if you cannot find it in the index perhaps it did not happen.

months were compared. This resulted in the analysis of twenty-six variable sets (see table 5.1).

There are only three variables which are significant at the .05 level, and an additional three are significant at the .10 level. At the start of the war interruption only the duration of economic violence shows a significant (.10) change. The intercept of these variables increases from 1 to 4 days per month. The duration of "violent clashes" and "all events" intercepts, while not statistically significant, increases from 2 to 5.3 days per month, with no appreciable slope changes. The number of deaths for violent clashes and all events also increases in the postwar months, but deaths related to economic violence do not. The third measure of intensity, arrests, shows no measurable slope or intercept changes.

At the end of the war the duration of economic violence, clashes, and all events show increases in intercept level significant at the

TABLE 5.1.
SPANISH-AMERICAN WAR

| | Slope | | Intercept | | Single | Walker-Lev | | Double- |
	Pre	Post	Pre	Post	Mood	Test 1	Test 3	Mood
			Start					
Duration								
All events	.01	−.01	1.95	5.32	.88	1.71	.85	.86
Violent clashes	.01	−.01	1.73	5.38	.92	2.57	.84	.84
Deaths								
All events	.03	.00	.15	3.34	.71	.58	.00	.02
Violent clashes	.03	−.01	.15	4.08	.71	.88	.00	.01
Arrests								
All events	−.02	.00	1.81	.33	.10	2.14	.70	.91
Violent clashes	−.01	.00	1.22	.21	.07	1.24	.47	.74
			End					
Duration								
All events	.01	−.02	2.19	7.18	.49	2.39	3.21	1.85*
Violent clashes	.01	−.03	1.98	7.15	.51	3.38	3.09	1.83*
Deaths								
All events	.02	−.01	.34	4.47	.66	.62	.05	.25
Violent clashes	.02	−.02	.34	5.25	.66	.93	.06	.27
Arrests								
All events	−.02	.00	1.78	.61	.11	1.68	1.24	1.07
Violent clashes	−.01	.00	1.19	.44	.08	.96	.85	.89
			Deleted					
Duration								
All events	.01	−.02	1.95	6.72	.60	2.99	1.83	1.33†
Violent clashes	.01	−.02	1.83	6.71	.58	3.57	2.15	1.44†

TABLE 5.1 (*Continued*)

Start

Duration								
Economic	.01	−.01	.91	3.83	.53	1.11	2.52	1.53†
Social	.01	.00	.90	1.29	.86	1.07	.88	.99
Political	.00	.00	.11	.11	.21	.06	.18	.43
Deaths								
Economic	.02	.00	−.48	.69	.38	1.19	.16	.45
Social	.01	−.01	.37	2.66	.80	.23	.04	.16
Political	None recorded after 29th month							
Arrests								
Economic	−.02	.00	1.26	.33	.07	1.20	.58	.82
Social	None recorded							
Political	None recorded after 23rd month							

End

Duration								
Economic	.00	−.02	1.00	5.11	.11	1.91	4.81*	2.24*
Social	.00	−.01	1.03	1.80	.74	.88	.04	.16
Political	.00	.00	.12	.17	.17	.02	.74	.85
Deaths								
Economic	.01	.00	−.37	1.10	.34	1.09	.02	.09
Social	.01	−.01	.46	3.38	.76	.28	.09	.32
Political	None recorded after 29th month							
Arrests								
Economic	−.01	.00	1.23	.61	.08	.89	1.02	.98
Social	None recorded							
Political	None recorded after 23rd month							

* Significant at the .05 level.
† Significant at the .10 level.

.05 level, with slopes remaining constant. Deaths, while again not statistically significant, increase for both violent clashes and all events; in addition, socially related deaths increase in the postwar period, with slopes remaining constant. This indicates a step-level increase in violence. Arrests remain constant at a low level throughout the period.

It is expected that by deleting the war months the results should follow the pattern of the start- and end-of-war interruptions. While slopes remain constant in the pre- and postwar periods, the intercepts rise for duration of all events (.10) and violent clashes (.10) by almost five days per month. In conclusion, there is a step-level increase following the initiation of war and in the postwar months. Duration of events is the most affected indicator although the same pattern is exhibited by deaths. Violence related to economics appears to account for the increase in duration, but apparently not for the increase in deaths.

WORLD WAR I

The impact of World War I was evaluated on forty-five [4] variable sets (see table 5.2). Twenty-six of these variables showed significant changes at the .05 level, and an additional three at the .10 level. The U.S. entered World War I as an official combatant in April, 1917. The number of days per month during which violence took place shows significant drops in level at that time for all categories of violence with the exceptions of political and government-initiated violence, which had sharp increases, and antigovernment-initiated violence, which evidenced no slope or intercept differences. There were a number of statistically significant slope changes, but the magnitude of those changes was very small for the variables which showed significant intercept differences. Recorded deaths, with the exception of àn increase (.05) from zero to one for political deaths, remained constant. The number of arrests in this period rose appreciably, with only the social and antigovernment categories showing no change. Government-initiated and political arrests, as well as arrests in the total events category, showed an intercept increase of over fifty-five per month. In addition, there were four significant slope differences (all events, government-initiated, economic, and political). It is significant that while the number of violent events decreased in these months, political and government-initiated duration and arrests rose markedly after the involvement in war.

The end of the war in November, 1918, after nineteen months of American participation, evidenced significant (.05) rises of duration in all categories of violent events. Significant slope changes also occurred in every category. Recorded deaths rose for all but government-initiated and economic events; only politically related deaths evidenced any slope change, and that was negative. The most marked change occurred in the recorded arrests category with significantly (.05) increased intercepts, from less than 1 to 105 in the all events category. Antigovernment-initiated, government-initiated, and political events sharply increased,

---

4. The increase after the Spanish-American War of the number of events to be evaluated was due to the number of government-initiated and antigovernment-initiated violent incidents, which allowed those categories to be included in the World War I analyses. There were only sixteen recorded government-initiated events in the 1890–1906 period, and none between 1894 and 1900 (after the Homestead and Pullman strikes). There were twenty-one recorded antigovernment-initiated events, with most occurring after 1901. For both of these categories there were many more observation points than events, which prevented statistical analysis.

TABLE 5.2
WORLD WAR I

	Slope		Intercept		Single-Mood	Walker-Lev		Double-Mood
	Pre	Post	Pre	Post		Test 1	Test 3	
Duration			*Start*					
All	−.14	.04	6.63	5.26	.92	6.28*	6.04*	3.25*
Clashes	−.11	.05	5.34	1.74	.49	7.49*	2.12	2.39*
Government-initiated	−.02	−.02	1.05	3.36	2.22*	.04	17.77*	3.98*
Antigovernment-initiated	.00	.01	.24	.30	.66	.16	.40	.73
Deaths								
All	−.07	.07	2.64	−2.02	.76	1.79	.05	.70
Clashes	−.06	.07	2.43	−2.67	.76	1.86	.00	.53
Government-initiated	.00	.00	.00	.15	I	.00	1.04	.92
Antigovernment-initiated	−.04	.00	.21	.55	.02	.02	.30	.55
Arrests								
All	−.43	−.53	15.77	78.49	4.10*	.04	13.74*	3.36*
Clashes	−.19	−.04	7.00	8.58	.25	.86	2.18	1.71*
Government-initiated	−.23	−.46	8.77	66.25	6.57*	.20	10.34*	2.82*
Antigovernment-initiated	.00	−.03	.00	3.72	I	.12	1.00	.80
Duration			*End*					
All	−.04	−.09	4.74	19.69	.05	.93	24.11*	4.99*
Clashes	−.04	−.02	3.98	9.65	.00	.13	13.91*	3.64*
Government-initiated	.01	−.06	.53	7.56	.10	15.74*	14.16*	4.45*
Antigovernment-initiated	.00	−.02	.24	2.76	.61	1.59	7.48*	2.88*

TABLE 5.2 (Continued)

	Slope		Intercept		Single-Mood	Walker-Lev		Double-Mood
	Pre	Post	Pre	Post		Test 1	Test 3	
Deaths								
All	−.03	.01	1.87	4.90	.12	.16	3.02	1.67*
Clashes	−.03	.03	1.73	2.05	.15	.42	2.04	1.33†
Government-initiated	.00	.00	−.04	−.15	.46	.00	.97	.98
Antigovernment-initiated	.00	−.02	.19	3.00	.09	.62	2.02	1.51†
Arrests								
All	.40	−.77	−.09	104.93	.63	6.61*	.81	1.25
Clashes	−.03	−.06	3.92	10.36	.13	.05	.62	.80
Government-initiated	.43	−.60	−4.01	81.59	.65	5.63*	.19	.74
Antigovernment-initiated	.06	−.11	.00	13.13	I	2.50	2.65	1.83*
			Start					
Duration								
Economic	−.14	.06	6.03	−1.54	.31	11.45*	.00	1.24
Social	−.01	.02	.40	1.23	1.64*	.73	4.19*	2.22*
Political	.00	−.04	.21	5.14	3.00*	3.06	28.01*	4.32*
Deaths								
Economic	−.06	.04	2.26	−2.55	.36	3.22	.19	.25
Social	−.01	.03	.38	−.75	.09	.26	.04	.37
Political	.00	−.01	.00	.98	R	1.50	9.30*	2.39*
Arrests								
Economic	−.43	−.16	15.70	19.79	.44	2.13	6.43*	2.90*
Social	.00	.03	.00	−.31	R	.06	.04	.28
Political	.00	−.36	.07	55.41	177.19*	.64	8.31*	2.39*

Duration								
Economic	−.09	−.01	5.05	6.00	.43	2.44	13.72*	3.49*
Social	.02	−.02	.01	4.62	.74	1.61	7.18*	2.83*
Political	.03	−.06	−.34	7.47	.00	22.47*	1.84	2.07*
Deaths								
Economic	−.03	.05	1.72	−2.87	.32	2.48	.58	.55
Social	.00	−.02	.26	4.65	.08	.03	1.65	1.29†
Political	.01	−.01	−.11	1.18	.67	9.18*	.05	.57
Arrests								
Economic	−.05	−.11	8.11	14.01	.20	.15	.03	.23
Social	.00	−.02	−.02	5.16	.27	.06	.79	.91
Political	.45	−.52	−8.18	72.90	.71	5.83*	.20	.76
Deleted								
Duration								
All	−.14	−.08	6.63	16.63	1.20	.56	27.41*	5.28*
Clashes	.11	−.02	5.34	8.50	1.11	2.14	13.05*	3.82*
Government-initiated	−.02	−.05	1.05	6.12	1.17	2.25	32.15*	5.37*
Antigovernment-initiated	.00	−.01	.23	2.23	.66	.64	6.05*	2.29*

*Significant at the .05 level.
†Significant at the .10 level.

accounting for most of the change, but violent clashes and economic and social arrests, while not statistically significant, also increased. The end of the war thus brought clear step-level increases in all categories of violence recorded, with government-initiated and political events demonstrating the greatest changes.

Once again the war months were deleted to compare pre- and postwar months for changes in the duration of violence. The four categories tested (all events, government-initiated, violent clashes, and antigovernment-initiated) displayed significant (.05) increases in intercept and, unlike the Spanish-American War events, significant negative slope changes as well, although none of the slope changes was of any appreciable magnitude.

The two wars, while not evidencing comparability over all events, do show a number of important similarities and step changes. At the start of both wars there were significant intercept changes, although, with the exception of government-initiated violence and political violence (duration), the direction was negative in the World War I period and positive in the Spanish-American War period. Both duration levels and deaths increase in the statistically significant categories in the two postwar periods, while arrests increase after World War I and no significant changes occur after the Spanish-American War. It should be reiterated that the arrests were mainly attributed to government-initiated and politically violent events in the post–World War I period and that there was not enough recorded government-initiated violence to analyze in the post–Spanish-American War period. When both war periods were removed for the purpose of comparison of pre- and postwar months only, there were significant step-level increases in the number of days per month of violence in both postwar periods.

DISCUSSION

The hypothesis that the post-event distribution of domestic violence after both wars was part of either a seasonal or a longer-term trend does not appear tenable, given the fact that the prewar observations do not exhibit the same trend. It is clearly not the case that pre-event values can be extrapolated to obtain the results evidenced. The postwar behavior appears to be part of a different trend — that is, sensitive to different forces — than the prewar behavior.

Three substantive hypotheses were explicated earlier predicting changes in the postwar interactions and behavior. While the quasi-experimental approach does not allow inferences to be drawn in terms of strength of relationship, changes in the pattern of violence in terms of these hypotheses may be indicated.

Hypothesis 1: Wartime economic mobilization brings new groups into the productive process and enhances the positions of groups relative to the dominant segments, thus intensifying economic conflict and violence.

After both the start and end of the Spanish-American War, the number of days of economic violence increased. However, after the start of World War I economic violence fell, although it rose again after the Armistice. Link (1966:309) points out that the average number of strikes during the period 1916–21 was 3,503 per year and that during 1919 there were 2,665 strikes involving more than four million workers (Link 1966:234). However, despite the growth of strikes, the level of violence during World War I was low, and the violence was mainly directed against strikers. The increase in violence after 1919 may be explained by the twin factors of the government's policy after the war of sending troops to trouble spots as a precautionary measure and by the fact that employers who had accepted the tremendous union growth (from 2,772,000 to 4,881,000 during 1916–21) as a wartime necessity (or as a government fiat) were now anxious to rid themselves of labor organizations (Taft and Ross 1969:332–33).

Hypothesis 2: Wartime social mobility increases the status positions of underdog social groups relative to the dominant segments, which increases hostilities between them.

Both the initiation of the Spanish-American War and the Armistice saw no change in social violence. However, the initiation of World War I was accompanied by increases in violence. It should be stressed that the last sixteen years of the nineteenth century witnessed more than 2,500 lynchings of blacks, and from 1900 to 1914 there were an additional 1,100 black victims (Link 1966:31). It also should be noted that the period before the Spanish-American War witnessed very high levels of nativist violence; while these did not increase in the post–Spanish-American War period, nativists did once again display "symptoms of hysteria and violence" by 1914 (Higham 1967:183). Higham posits that the acquisition of an empire after the Spanish-American War fortified the confidence of dominant segments renewed by prosperity after the war, by relief from class conflict, and by a psychologically invigorating war, and that this accounted for the decline of nativism until 1914 (Higham 1967:108).

Hypothesis 3: The economic and social changes of the war generate demands for the reallocation of political power and rewards, which intensifies conflict and violence. In addition, success in war increases the power and prestige of dominant groups and intensifies efforts to maintain control of the political system.

As indicated earlier, there were no changes in political violence after the Spanish-American War and not enough recorded government-initiated violent events for analysis. During and after World War I the picture is quite different, with a sharp increase in both government-initiated and political violence, as well as a small increase in antigovernment-initiated violence at the end of the war. The great increase in violence can be traced to the legislative and ideological foundations laid during the war for a nationalistic, antiradical crusade (the Espionage Act of 1917, the Sedition Act of 1918, and the congressional authorization of the Justice and Labor Department to deport any alien simply on grounds of belonging to an organization which advocated revolt or sabotage in 1918). No such bills were passed during the Spanish-American War, perhaps because of the swift success of the war effort. As a result, over 2,100 persons were prosecuted under the alien and sedition acts, and only ten of them for actual sabotage (Link 1966:212). In addition, the Red Scare of 1919–20 and the resultant Palmer raids led to the arrest of over 4,000 persons, including bona fide citizens, most of whom were eventually released after spending time in government stockades. However, 556 aliens were deported simply for being Communist Party members. Further in-depth research is needed to analyze the forces that resulted in the postwar preparations for return to "normalcy."

In conclusion, it should be stressed that the quasi-experimental approach located step-level changes in violence in the periods following both wars. The importance of comparing pre- and postwar years for changes in violence due to involvement in war is highlighted by comparison with the results of a study by Sheldon Levy:

The absolute number of politically violent events has been rising throughout American history with the exception of three periods. One was in the decade prior to the turn of the century. The second was prior to and following World War One. This was followed by a sharp rise during the depression period but there was another drop shortly before, through and shortly after the second World War. (1969:83)

Thus, these increases in postwar violence appear to occur during two periods which are not characterized by long-term increases in violence. An approach which did not have the ability to distinguish short-term slope and intercept changes would likely miss a substantial change within these long-term trends.

PART THREE

EVALUATING POLICY

The need for research on policy questions should be clear. Decisions are made and massive expenditures undertaken by governments. Better understanding of the reasons for success and failure of government projects would seem essential. Yet, as one economist (Haveman 1972:1) has noted, "Only very recently has it been possible to find any significant research at all that focuses on the economic results of public undertakings after they have had time to develop a performance record." Often analysts have not been able to define what constitutes "success" and "failure." Research designs have been weak; the record has been mixed.

Such problems have led to a call for social experimentation. As Schultze et al. (1972) have noted, "In many areas of social policy, no one really knows which techniques or approaches are successful and which are not." Although quasi-experiments cannot substitute for true experiments, they provide a highly useful tool for the policy analyst. This tool should not be taken as a substitute for benefit-cost analysis, but rather as a valuable adjunct to economic-based approaches.

The logic of quasi-experimentation can be compared with that of benefit-cost analysis, the dominant method used for the analysis of public expenditures in the late 1960s.

> Benefit-cost analysis seeks to rank alternative uses of public funds. In the usual situation, several projects are competing for scarce funds. Each one can be pursued at one or more scales. The problem may be to choose which of several projects to engage in and then to choose its optimal scale from the several which may be available. Another way to see the problem is: choose the best scale for each project and then choose the best alternative among several projects, each at its optimal scale. For either formulation, we seek a ranking function with which to compare alternatives. (Merewitz and Sosnick 1971:85)

The quasi-experimental approach is based on a different conceptual framework and confronts different problems than does benefit-cost analysis. Some of these differences are outlined in the accompanying table.

BENEFIT-COST ANALYSIS AND QUASI-EXPERIMENTATION

	Benefit-Cost Analysis	Quasi-Experimentation
Disciplinary antecedents	Economics	Psychology
Theoretical emphases	Production functions, economic efficiency	More varied, both theoretical and atheoretical
Presumptions as to knowledge base	Great deal of knowledge assumed	Little knowledge assumed
Faith in analysis	Considerable	Slight
Time frame for study	Before initiation of project (*ex ante* orientation)	Before and after project is initiated (*ex ante, ex post* orientation)
Time period before recommendations are produced	Relatively short	Relatively long
Where used in past	Particularly public works and defense programs	Some use with social programs
Variables emphasized	Monetized costs and benefits	Almost any
Dominant concerns	Concern with analytical model	Concern with strength of research design
Central focus	Focus on the problem or project	Focus on match between design and problem
Problems needing more work in many analyses	Externalities	Selection

Benefit-cost and quasi-experimental approaches differ markedly in their faith in the intellectual power of *ex ante* analysis. Coming out of economics—the social science with the most well-developed theoretical paradigm—much of benefit-cost analysis has proceeded on the implicit assumption that *ex post* analysis was unnecessary. Quasi-experimental and other evaluation approaches have been developed out of theoretically less-mature behavioral sciences.

Given weak theory, the emphasis has been on research design to try to establish "what affects what." Data and strong designs both provide the discipline needed to effectively test good hypotheses and are necessary correctives to ideas and hypotheses which may be very wrong.

In comparing benefit-cost analysis and quasi-experimentation, questions of time perspective seem central. Benefit-cost analysis has emphasized the evaluation of particular projects, one at a time. Like cross-sectional research, benefit-cost analysis characteristically promises results (better policies) rather quickly. An experimental or quasi-experimental approach cannot make such promises. The question of living up to the promises – of accuracy of results – is, of course, salient in this connection. Moreover, better long-term results might come from a consideration of research design at the time project proposals are rated. Several projects might mesh together and permit testing assumptions and effects in a more definitive fashion.

Problems with benefit-cost analysis are receiving increased attention. Haveman stresses these in his valuable research comparing the *ex ante* analysis of public sector investments with their *ex post* performance:

> unless procedures are constantly revised on the basis of performance feedback from existing undertakings, the credibility of ex ante analysis will, and should, be challenged. The serious discrepancies between projected and realized costs and benefits described in this study do little to instill confidence in current ex ante analysis. (1972:111)

Such discrepancies are particularly noteworthy because they are found in water resources research, a substantive area in which such analytical techniques have been applied longer than they have in almost any other area.

Benefit-cost analysis does, however, promise substantial improvement over the way many government organizations have spent their funds much of the time. *Ex ante* benefit-cost analysis demands the collection of considerable data before major expenditures are undertaken. In the social arena, data have often not been collected at the "before" stage; evaluation researchers are frequently called in *after* a project is well under way or almost completed. Without adequate *before* data, researchers must rely on weak designs, the statistical "matching" of participants and nonparticipants, and so forth. Administrative concern with both analysis and design would lead to a merging of approaches and better policy evaluation – if not better policy.

Given the emphasis on choice and comparison implicit in benefit-cost analysis, this technique is particularly appropriate for providing advice relevant to policy makers (even if the recommendations are inappropriate or incorrect). When applied to policy problems, quasi-experimental methods are oriented toward evaluating a particular program. Comparisons are generally of the before-after type, or between an experimental and a control group. There is no reason why quasi-experimental techniques could not be used to explicitly compare two different policies, but few such comparisons have been made in the past. The articles included here discuss such quasi-experimental designs and their policy applications. Campbell's piece (chapter 6) treats a number of these designs, providing an important overview of this area. True experiments and quasi-experiments are contrasted, while administrators are urged to experiment.

The selection by Shingles (chapter 7) discusses some of the unintended consequences of a federal policy initiative – the Head Start program. While not strictly policy analysis, Shingles' essay integrates a policy concern with an interest in theory – the mass society and civic culture theses. In this paper a pretest-posttest design is used, with participation as the intervening event and attitudinal measures of self- and system-evaluation as the dependent variables.

The selection by Roos and Bohner (chapter 8) applies the interrupted time-series design to a relatively new area, that of environmental protection. This essay builds on earlier work (Caldwell with Roos 1971), extending the research design and suggesting ways to evaluate a number of pollution abatement strategies.

Noralou Roos (chapter 9) draws on her experience as a short-term bureaucrat to assess the possibilities for greater use of quasi-experimental designs in evaluation research. Some suggestions are made for increasing the acceptance of more complex, but stronger, research designs. Her paper mentions one problem shared by both benefit-cost analysis and quasi-experimental research. This difficulty – the biases of administrative agencies – appears endemic to all efforts at systematic analysis. As Merewitz and Sosnick put it:

> Agencies often perform benefit-cost analyses on projects they themselves will execute if they can demonstrate feasibility. . . . Such situations lead to possible conflicts of interest. (1971:276)

Campbell (chapter 6 of this volume) has noted with regard to such conflict of interest:

> Most administrators wisely prefer to limit the evaluations to those the outcomes of which they can control, particularly insofar as published outcomes or press releases are concerned.

Finally, some of the problems of institutionalizing evaluation efforts in government have been dealt with in the paper by Noralou Roos. Although the use of systematic evaluation appears to be increasing, there are still numerous problems connected with classification and suppression of results, retaliation against employees reporting "unfavorable" findings, and so forth. Perhaps the 1970s will see a more enthusiastic use of experimentation and quasi-experimentation, followed by a winnowing out of the more inappropriate procedures and weaker designs. Such a course might be desirable and, to some degree, would be in line with the history of programming-planning-budgeting (Merewitz and Sosnick 1971).

6

REFORMS AS EXPERIMENTS

DONALD T. CAMPBELL

The United States and other modern nations should be ready for an experimental approach to social reform, an approach in which we try out new programs designed to cure specific social problems, in which we learn whether or not these programs are effective, and in which we retain, imitate, modify, or discard them on the basis of apparent effectiveness on the multiple imperfect criteria available. Our readiness for this stage is indicated by the inclusion of specific provisions for program evaluation in the first wave of the "Great Society" legislation, and by the current congressional proposals for establishing "social indicators" and socially relevant "data banks." So long have we had good intentions in this regard that many may feel we are already at this stage, that we already are continuing or discontinuing programs on the basis of assessed effectiveness. It is a theme of this article that this is not at all so, that most ameliorative programs end up with *no* interpretable evaluation (Etzioni 1968; Hyman and Wright 1967; Schwartz 1961). We must look hard at the sources of this condition, and design ways of overcoming the difficulties. This article is a preliminary effort in this regard.

Many of the difficulties lie in the intransigencies of the research

This article originally appeared in *The American Psychologist*, Volume 24, number 4 (April, 1969), pp. 409–29, copyright 1969 by the American Psychological Association. It is reprinted with minor revisions by permission of the publisher. The preparation of the paper was supported by National Science Foundation Grant GS1309X. Versions of the paper have been presented at the Northwestern University Alumni Fund Lecture, January 24, 1968; to the Social Psychology Section of the British Psychological Society at Oxford, September 20, 1968; to the International Conference on Social Psychology at Prague, October 7, 1968 (under a different title); and to several other groups.

setting and in the presence of recurrent seductive pitfalls of interpretation. The bulk of this article will be devoted to these problems. But the few available solutions turn out to depend upon correct administrative decisions in the initiation and execution of the program. These decisions are made in a political arena, and involve political jeopardies that are often sufficient to explain the lack of hardheaded evaluation of effects. Removing reform administrators from the political spotlight seems both highly unlikely, and undesirable even if it were possible. What is instead essential is that the social scientist research advisor understand the political realities of the situation, and that he aid by helping create a public demand for hardheaded evaluation, by contributing to those political inventions that reduce the liability of honest evaluation, and by educating future administrators to the problems and possibilities.

For this reason, there is also an attempt in this article to consider the political setting of program evaluation, and to offer suggestions as to political postures that might further a truly experimental approach to social reform. Although such considerations will be distributed as a minor theme throughout this article, it seems convenient to begin with some general points of this political nature.

POLITICAL VULNERABILITY FROM KNOWING OUTCOMES

It is one of the most characteristic aspects of the present situation that *specific reforms are advocated as though they were certain to be successful.* For this reason, knowing outcomes has immediate political implications. Given the inherent difficulty of making significant improvements by the means usually provided and given the discrepancy between promise and possibility, most administrators wisely prefer to limit the evaluations to those the outcomes of which they can control, particularly insofar as published outcomes or press releases are concerned. Ambiguity, lack of truly comparable comparison bases, and lack of concrete evidence all work to increase the administrator's control over what gets said, or at least to reduce the bite of criticism in the case of actual failure. There is safety under the cloak of ignorance. Over and above this tie-in of advocacy and administration, there is another source of vulnerability in that the facts relevant to experimental program evaluation are also available to argue the general efficiency and honesty of administrators. The public availability of such facts reduces the privacy and security of at least some administrators.

Even where there are ideological commitments to a hardheaded evaluation of organizational efficiency, or to a scientific organization of society, these two jeopardies lead to the failure to evaluate organizational experiments realistically. If the political and administrative system has committed itself in advance to the correctness and efficacy of its reforms, it cannot tolerate learning of failure. To be truly scientific we must be able to experiment. We must be able to advocate without that excess of commitment that blinds us to reality testing.

This predicament, abetted by public apathy and by deliberate corruption, may prove in the long run to permanently preclude a truly experimental approach to social amelioration. But our needs and our hopes for a better society demand we make the effort. There are a few signs of hope. In the United States we have been able to achieve cost-of-living and unemployment indices that, however imperfect, have embarrassed the administrations that published them. We are able to conduct censuses that reduce the number of representatives a state has in Congress. These are grounds for optimism, although the corrupt tardiness of state governments in following their own constitutions in revising legislative districts illustrates the problem.

One simple shift in political posture which would reduce the problem is the shift from the advocacy of a specific reform to the advocacy of the seriousness of the problem, and hence to the advocacy of persistence in alternative reform efforts should the first one fail. The political stance would become: "This is a serious problem. We propose to initiate policy A on an experimental basis. If after five years there has been no significant improvement, we will shift to policy B." By making explicit that a given problem solution was only one of several that the administrator or party could in good conscience advocate, and by having ready a plausible alternative, the administrator could afford honest evaluation of outcomes. Negative results, a failure of the first program, would not jeopardize his job, for his job would be to keep after the problem until something was found that worked.

Coupled with this should be a general moratorium on ad hominem evaluative research—that is, on research designed to evaluate specific administrators rather than alternative policies. If we worry about the invasion of privacy problem in the data banks and social indicators of the future (e.g., Sawyer and Schechter 1968), the touchiest point is the privacy of administrators. If we threaten this, the measurement system will surely be sabotaged in the innumerable ways possible. While this may sound unduly pessimistic, the recurrent anecdotes of administrators attempting to squelch unwanted research findings convince me of its accuracy. But we should be able to evaluate those alternative policies that a given administrator has the option of implementing.

FIELD EXPERIMENTS
AND QUASI-EXPERIMENTAL DESIGNS

In efforts to extend the logic of laboratory experimentation into the "field," and into settings not fully experimental, an inventory of threats to experimental validity has been assembled, in terms of which some fifteen or twenty experimental and quasi-experimental designs have been evaluated (Campbell 1957, 1963; Campbell and Stanley 1963). In the present article only three or four designs will be examined, and therefore not all of the validity threats will be relevant, but it will provide useful background to look briefly at them all. Following are nine threats to internal validity: [1]

1. *History:* events, other than the experimental treatment, occurring between pretest and posttest and thus providing alternate explanations of effects.
2. *Maturation:* processes within the respondents or observed social units producing changes as a function of the passage of time per se, such as growth, fatigue, secular trends, etc.
3. *Instability:* unreliability of measures, fluctuations in sampling persons or components, autonomous instability of repeated or "equivalent" measures. (This is the only threat to which statistical tests of significance are relevant.)

---

1. This list has been expanded from the major previous presentations by the addition of *Instability* (but see Campbell 1968; Campbell and Ross, 1968). This has been done in reaction to the sociological discussion of the use of tests of significance in nonexperimental or quasi-experimental research (e.g., Selvin [1957]; and as reviewed by Galtung [1967:358–89]). On the one hand, I join with the critics in criticizing the exaggerated status of "statistically significant differences" in establishing convictions of validity. Statistical tests are relevant to, at best, one out of fifteen or so threats to validity. On the other hand, I join with those who defend their use in situations where randomization has not been employed. Even in those situations, it is relevant to say or to deny, "This is a trivial difference. It is of the order that would have occurred frequently *had* these measures been assigned to these classes solely by chance." Tests of significance, making use of random reassignments of the actual scores, are particularly useful in communicating this point.

4. *Testing:* the effect of taking a test upon the scores of a second testing; the effect of publication of a social indicator upon subsequent readings of that indicator.

5. *Instrumentation:* in which changes in the calibration of a measuring instrument or changes in the observers or scores used may produce changes in the obtained measurements.

6. *Regression artifacts:* pseudo-shifts occurring when persons or treatment units have been selected upon the basis of their extreme scores.

7. *Selection:* biases resulting from differential recruitment of comparison groups, producing different mean levels on the measure of effects.

8. *Experimental mortality:* the differential loss of respondents from comparison groups.

9. *Selection-maturation interaction:* selection biases resulting in differential rates of "maturation" or autonomous change.

If a change or difference occurs, these are rival explanations that could be used to explain away an effect and thus to deny that in this specific experiment any genuine effect of the experimental treatment had been demonstrated. These are faults that true experiments avoid, primarily through the use of randomization and control groups. In the approach here advocated, this checklist is used to evaluate specific quasi-experimental designs. This is evaluation, not rejection, for it often turns out that for a specific design in a specific setting the threat is implausible, or that there are supplementary data that can help rule it out even where randomization is impossible. The general ethic, here advocated for public administrators as well as social scientists, is to use the very best method possible, aiming at "true experiments" with random control groups. But where randomized treatments are not possible, a self-critical use of quasi-experimental designs is advocated. We must do the best we can with what is available to us.

Our posture vis-à-vis perfectionist critics from laboratory experimentation is more militant than this: the only threats to validity that we will allow to invalidate an experiment are those that admit of the status of empirical laws more dependable and more plausible than the law involving the treatment. The mere possibility of some alternative explanation is not enough — it is only the *plausible* rival hypotheses that are invalidating. Vis-à-vis correlational studies and common-sense descriptive studies, on the other hand, our stance is one of greater conservatism. For example, because of the specific methodological trap of regression artifacts, the sociological tradition of "ex post facto" designs (Chapin 1947; Greenwood 1945) is totally rejected (Campbell and Stanley 1963:240–41, 1966:70–71).

Threats to external validity, which follow, cover the validity problems involved in interpreting experimental results – the threats to valid generalization of the results to other settings, to other versions of the treatment, or to other measures of the effect: [2]

1. *Interaction effects of testing:* the effect of a pretest in increasing or decreasing the respondent's sensitivity or responsiveness to the experimental variable, thus making the results obtained for a pretested population unrepresentative of the effects of the experimental variable for the unpretested universe from which the experimental respondents were selected.
2. *Interaction of selection and experimental treatment:* unrepresentative responsiveness of the treated population.
3. *Reactive effects of experimental arrangements:* "artificiality"; conditions making the experimental setting atypical of conditions of regular application of the treatment ("Hawthorne effects").
4. *Multiple-treatment interference:* where multiple treatments are jointly applied, effects atypical of the separate application of the treatments.
5. *Irrelevant responsiveness of measures:* all measures are complex, and all include irrelevant components that may produce apparent effects.
6. *Irrelevant replicability of treatments:* treatments are complex, and replications of them may fail to include those components actually responsible for the effects.

These threats apply equally to true experiments and quasi-experiments. They are particularly relevant to applied experimentation. In the cumulative history of our methodology, this class of threats was first noted as a critique of true experiments involving pretests (Schanck and Goodman 1939; Solomon 1949). Such experiments provided a sound basis for generalizing to other *pretested* populations, but the reactions of unpretested populations to the treatment might well be quite different. As a result, there has been an

---

2. This list has been lengthened from previous presentations to make more salient threats 5 and 6, which are particularly relevant to social experimentation. Discussion in previous presentations (Campbell 1957:309–10; Campbell and Stanley 1963:203–4) had covered these points, but they had not been included in the checklist.

advocacy of true experimental designs obviating the pretest (Campbell 1957; Schanck and Goodman 1939; Solomon 1949) and a search for nonreactive measures (Webb, Campbell, Schwartz, and Sechrest 1966).

These threats to validity will serve as a background against which we will discuss several research designs particularly appropriate for evaluating specific programs of social amelioration. These are the "interrupted time-series design," the "control-series design," the "regression discontinuity design," and various "true experiments." The order is from a weak but generally available design to stronger ones that require more administrative foresight and determination.

INTERRUPTED TIME-SERIES DESIGN

By and large, when a political unit initiates a reform it is put into effect across the board, with the total unit being affected. In this setting the only comparison base is the record of previous years. The usual mode of utilization is a casual version of a very weak quasi-experimental design, the one-group pretest-posttest design.

A convenient illustration comes from the 1955 Connecticut crackdown on speeding, which sociologist H. Laurence Ross and I have been analyzing as a methodological illustration (Campbell and Ross 1968; Glass 1968; Ross and Campbell 1968). After a record high of traffic fatalities in 1955, Governor Abraham Ribicoff instituted an unprecedentedly severe crackdown on speeding. At the end of a year of such enforcement there had been but 284 traffic deaths as compared with 324 the year before. In announcing this the governor stated, "With the saving of 40 lives in 1956, a reduction of 12.3 per cent from the 1955 motor vehicle death toll, we can say that the program is definitely worthwhile." These results are graphed in figure 6.1, with a deliberate effort to make them look impressive.

In what follows, while we in the end decide that the crackdown had some beneficial effects, we criticize Ribicoff's interpretation of his results, from the point of view of the social scientist's proper standards of evidence. Were the now Senator Ribicoff not the man of stature that he is, this would be most unpolitic, because we could be alienating one of the strongest proponents of social experimentation in our nation. Given his character, however, we may feel sure that he shares our interests both in a progressive program of experimental social amelioration, and in making the most hard-

FIGURE 6.1.   Connecticut Traffic Fatalities

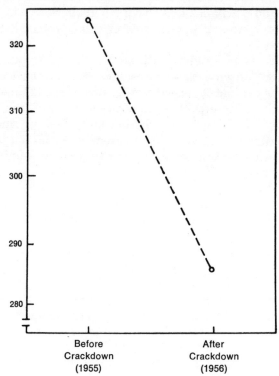

headed evaluation possible of these experiments. Indeed, it was his integrity in using every available means at his disposal as governor to make sure that the unpopular speeding crackdown was indeed enforced that make these data worth examining at all. But the potentials of this one illustration, and our political temptation to substitute for it a less touchy one, point to the political problems that must be faced in experimenting with social reform.

Keeping figure 6.1 and Ribicoff's statement in mind, let us look at the same data presented as a part of an extended time series in figure 6.2 and go over the relevant threats to internal validity.

HISTORY

Both presentations fail to control for the effects of other potential change agents. For instance, 1956 might have been a particularly dry year, with fewer accidents due to rain or snow. Or there might have been a dramatic increase in use of seat belts, or other safety

FIGURE 6.2.   Connecticut Traffic Fatalities (Same Data as in
Figure 6.1 Presented as Part of an Extended Time Series)

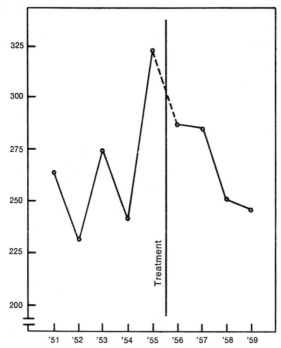

features. The advocated strategy in quasi-experimentation is not
to throw up one's hands and refuse to use the evidence because of
this lack of control, but rather to generate by informed criticism
appropriate to this specific setting as many *plausible* rival hypotheses as possible, and then to do the supplementary research, as into
weather records and safety-belt sales, for example, which would
reflect on these rival hypotheses.

MATURATION

This is a term coming from criticisms of training studies of children.
Applied here to the simple pretest-posttest data of figure 6.1, it
could be the plausible rival hypothesis that death rates were
steadily going down year after year (as indeed they are, relative to
miles driven or population of automobiles). Here the extended time
series has a strong methodological advantage, and rules out this
threat to validity. The general trend is inconsistently up prior to
the crackdown, and steadily down thereafter.

INSTABILITY

Seemingly implicit in the public pronouncement was the assumption that all of the change from 1955 to 1956 was due to the crackdown. There was no recognition of the fact that all time series are unstable even when no treatments are being applied. The degree of this normal instability is the crucial issue, and one of the main advantages of the extended time series is that it samples this instability. The great pretreatment instability now makes the treatment effect look relatively trivial. The 1955–56 shift is less than the gains of both 1954–55 and 1952–53. It is the largest drop in the series, but it exceeds the drops of 1951–52, 1953–54, and 1957–58 by trivial amounts. Thus the unexplained instabilities of the series are such as to make the 1955–56 drop understandable as more of the same. On the other hand, it is noteworthy that after the crackdown there are no year-to-year gains, and in this respect the character of the time series seems definitely to have changed.

The threat of instability is the only threat to which tests of significance are relevant. Box and Tiao (1965) have an elegant Bayesian model for the interrupted time series. Applied by Glass (1968) to our monthly data, with seasonal trends removed, it shows a statistically significant downward shift in the series after the crackdown. But as we shall see, an alternative explanation of at least part of this significant effect exists.

REGRESSION

In true experiments the treatment is applied independently of the prior state of the units. In natural experiments exposure to treatment is often a cosymptom of the treated group's condition. The treatment is apt to be an *effect* rather than, or in addition to being, a cause. Psychotherapy is such a cosymptom treatment, as is any other in which the treated group is self-selected or assigned on the basis of need. These all present special problems of interpretation, of which the present illustration provides one type.

The selection-regression plausible rival hypothesis works this way: Given that the fatality rate has some degree of unreliability, then a subsample selected for its extremity in 1955 would on the average, merely as a reflection of that unreliability, be less extreme in 1956. Has there been selection for extremity in applying this treatment? Probably so. Of all Connecticut fatality years, the most likely time for a crackdown would be after an exceptionally high year. If the time series showed instability, the subsequent year would on the average be less, or nearer to the general trend, *purely as a function of that instability.* Regression artifacts are probably the most recurrent form of self-deception in the experimental social reform literature. It is hard to make them intuitively obvious. Let us try again. Take any time series with variability, including one generated of pure error. Move along it as in a time dimension. Pick a point that is the "highest so far." Look then at the next point. On the average this next point will be lower, or nearer the general trend.

In our present setting the most striking shift in the whole series is the upward shift just prior to the crackdown. It is highly probable that this caused the crackdown, rather than, or in addition to, the crackdown causing the 1956 drop. At least part of the 1956 drop is an artifact of the 1955 extremity. While in principle the degree of expected regression can be computed from the auto-correlation of the series, we lack here a sufficiently extended body of data to do this with any confidence.

Advice to administrators who want to do genuine reality-testing must include attention to this problem, and it will be a very hard problem to surmount. The most general advice would be to work on chronic problems of a persistent urgency or extremity, rather than reacting to momentary extremes. The administrator should look at the pretreatment time series to judge whether or not instability plus momentary extremity will explain away his program gains. If it will, he should schedule the treatment for a year or two later, so that his decision is more independent of the one year's extremity. (The selection biases remaining under such a procedure need further examination.)

In giving advice to the *experimental* administrator, one is also inevitably giving advice to those *trapped* administrators whose political predicament requires a favorable outcome whether valid or not. To such trapped administrators the advice is to pick the very worst year, and the very worst social unit. If there is inherent instability, there is nowhere to go but up, for the average case at least.

TESTING

By testing we typically have in mind the condition under which a test of attitude, ability, or personality is itself a change agent, persuading, informing, practicing, or otherwise setting processes of change in action. No artificially introduced testing procedures are involved here. However, for the simple before-and-after design of figure 6.1, if the pretest were the first data collection of its kind ever publicized, this publicity in itself might produce a reduction in traffic deaths which would have taken place even without a speeding crackdown. Many traffic safety programs assume this. The longer time-series evidence reassures us on this only to the extent that we can assume that the figures had been published each year with equivalent emphasis.[3]

INSTRUMENTATION

Changes in instrumentation are not a likely flaw in this instance, but they would be if recording practices and institutional responsibility had shifted simultaneously with the crackdown. Probably in a case like this it is better to use raw frequencies rather than indices whose correction parameters are subject to periodic revision. Thus per capita rates are subject to periodic jumps as new census figures become available correcting old extrapolations. Analogously, a change in the miles per gallon assumed in estimating traffic mileage for mileage-based mortality rates might explain a shift. Such biases can of course work to disguise a true effect. Almost certainly, Ribicoff's crackdown reduced traffic speed (Campbell and Ross 1968). Such a decrease in speed increases the miles per gallon actually obtained, producing a concomitant drop in the estimate of miles driven, which would appear as an inflation of the estimate of mileage-based traffic fatalities if the same fixed approximation to actual miles per gallon were used, as it undoubtedly would be.

The "new broom" that introduces abrupt changes of policy is apt to reform the record keeping too, and thus confound reform treatments with instrumentation change. The ideal experimental administrator will, if possible, avoid doing this. He will prefer to keep comparable a partially imperfect measuring system rather than lose comparability altogether. The politics of the situation do not

3. No doubt the public and press shared the governor's special alarm over the 1955 death toll. This differential reaction could be seen as a negative feedback servosystem in which the dampening effect was proportional to the degree of upward deviation from the prior trend. Insofar as such alarm reduces traffic fatalities, it adds a negative component to the autocorrelation, increasing the regression effect. This component should probably be regarded as a rival cause or treatment rather than as artifact. (The regression effect is less as the positive autocorrelation is higher, and will be present to some degree insofar as this correlation is less than positive unity. Negative correlation in a time series would represent regression beyond the mean, in a way not quite analogous to negative correlation across persons. For an autocorrelation of lag 1, high negative correlation would be represented by a series that oscillated maximally from one extreme to the other.)

FIGURE 6.3. Number of Reported Larcenies under $50 in Chicago, Illinois, from 1942 to 1962 (Data from *Uniform Crime Reports for the United States*, 1942–62)

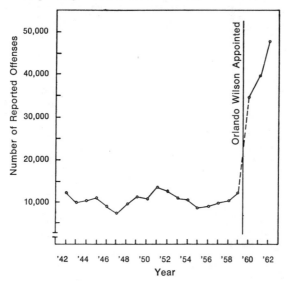

always make this possible, however. Consider, as an experimental reform, Orlando Wilson's reorganization of the police system in Chicago. Figure 6.3 shows his impact on petty larceny in Chicago— a striking *increase!* Wilson, of course, called this shot in advance, one aspect of his reform being a reform in the bookkeeping. (Note in the pre-Wilson records the suspicious absence of the expected upward secular trend.) In this situation Wilson had no choice. Had he left the record keeping as it was, for the purposes of better experimental design, his resentful patrolmen would have clobbered him with a crime wave by deliberately starting to record the many complaints that had not been getting into the books.[4]

Those who advocate the use of archival measures as social indicators (Bauer 1966; Gross 1966, 1967; Kaysen 1967; Webb et al. 1966) must face up not only to their high degree of chaotic error and systematic bias, but also to the politically motivated changes in record keeping that will follow upon their public use as social indicators (Etzioni and Lehman 1967). Not all measures are

4. Wilson's inconsistency in utilization of records and the political problem of relevant records are ably documented in Kamisar (1964). Etzioni (1968) reports that in New York City in 1965 a crime wave was proclaimed that turned out to be due to an unpublicized improvement in record keeping.

FIGURE 6.4. Number of Reported Murders and Nonnegligent Manslaughters in Chicago, Illinois, from 1942 to 1962 (Data from *Uniform Crime Reports for the United States*, 1942–62)

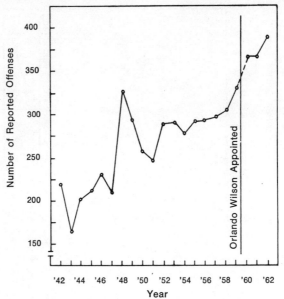

equally susceptible. In figure 6.4, Orlando Wilson's effect on homicides seems negligible one way or the other.

IRRELEVANT RESPONSIVENESS OF MEASURES

Of the threats to external validity, the one most relevant to social experimentation is *irrelevant responsiveness of measures*. This seems best discussed in terms of the problem of generalizing from indicator to indicator or in terms of the imperfect validity of all measures that is only to be overcome by the use of multiple measures of independent imperfection (Campbell and Fiske 1959; Webb et al. 1966).

For treatments on any given problem within any given governmental or business subunit, there will usually be something of a governmental monopoly on reform. Even though different divisions may optimally be trying different reforms, within each division there will usually be only one reform on a given problem going on at a time. But for measures of effect this need not and should not be the case. The administrative machinery should itself make multi-

FIGURE 6.5. Suspensions of Licenses for Speeding, as a Percentage of All Suspensions

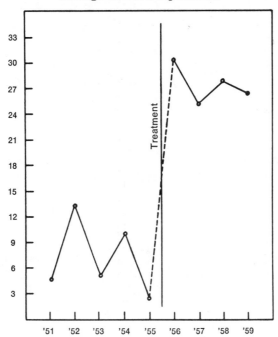

ple measures of potential benefits and of unwanted side effects. In addition, the loyal opposition should be allowed to add still other indicators, with the political process and adversary argument challenging both validity and relative importance, with social science methodologists testifying for both parties, and with the basic records kept public and under bipartisan audit (as are voting records under optimal conditions). This competitive scrutiny is indeed the main source of objectivity in sciences (Polanyi 1966, 1967; Popper 1963) and epitomizes an ideal of democratic practice in both judicial and legislative procedures.

The next few figures return again to the Connecticut crackdown on speeding and look to some other measures of effect. They are relevant to the confirming that there was indeed a crackdown, and to the issue of side effects. They also provide the methodological comfort of assuring us that in some cases the interrupted time-series design can provide clear-cut evidence of effect. Figure 6.5 shows the jump in suspensions of licenses for speeding — evidence that severe punishment was abruptly instituted. Again a note to

FIGURE 6.6.   Speeding Violations, as a Percentage of All Traffic Violations

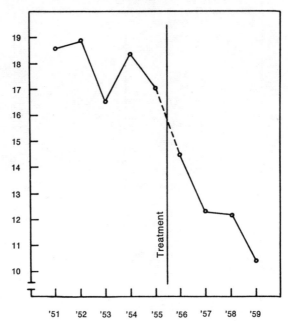

experimental administrators: with this weak design, *it is only abrupt and decisive changes that we have any chance of evaluating.* A gradually introduced reform will be indistinguishable from the background of secular change, from the net effect of the innumerable change agents continually impinging.

We would want intermediate evidence that traffic speed was modified. A sampling each year of a few hundred five-minute highway movies (random as to location and time) could have provided this at a moderate cost, but they were not collected. Of the public records available, perhaps the data of figure 6.6, showing a reduction in speeding violations, indicate a reduction in traffic speed. But the effects on the legal system were complex, and in part undesirable. Driving with a suspended license markedly increased (figure 6.7), at least in the biased sample of those arrested. Presumably because of the harshness of the punishment if guilty, judges may have become more lenient (figure 6.8), although this effect is of marginal significance.

The relevance of indicators for the social problems we wish to cure must be kept continually in focus. The social indicators

FIGURE 6.7. Arrested While Driving with a Suspended License, as a Percentage of Suspensions

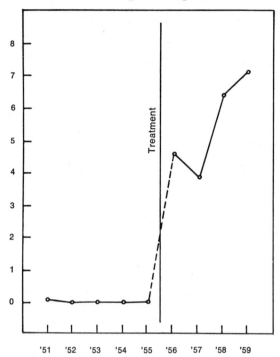

approach will tend to make the indicators themselves the goal of social action, rather than the social problems they but imperfectly indicate. There are apt to be tendencies to legislate changes in the indicators per se rather than changes in the social problems.

To illustrate the problem of the irrelevant responsiveness of measures, figure 6.9 shows a result of the 1900 change in divorce law in Germany. In a recent reanalysis of the data with the Box and Tiao (1965) statistic, Glass (Glass, Tiao, and Maguire 1971) has found the change highly significant, in contrast to earlier statistical analyses (Rheinstein 1959; Wolf, Lüke, and Hax 1959). But Rheinstein's emphasis would still be relevant: This indicator change indicates no likely improvement in marital harmony, or even in marital stability. Rather than reducing them, the legal change has made the divorce rate a less valid indicator of marital discord and separation than it had been earlier (see also Etzioni and Lehman 1967).

FIGURE 6.8. Percentage of Speeding Violations Judged Not Guilty

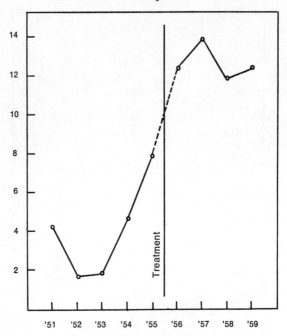

FIGURE 6.9. Divorce Rate for German Empire, 1881–1914

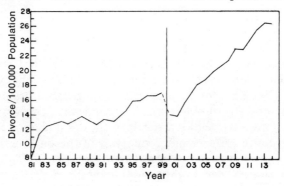

CONTROL-SERIES DESIGN

The interrupted time-series design as discussed so far is available for those settings in which no control group is possible, in which the total governmental unit has received the experimental treatment, the social reform measure. In the general program of quasi-experimental design, we argue the great advantage of un- treated comparison groups even where these cannot be assigned at random. The most common of such designs is the nonequivalent control group pretest-posttest design, in which for each of two natural groups, one of which receives the treatment, a pretest and posttest measure is taken. If the traditional mistaken practice is avoided of matching on pretest scores (with resultant regression artifacts), this design provides a useful control over those aspects of history, maturation, and test-retest effects shared by both groups. But it does not control for the plausible rival hypothesis of *selection-maturation interaction* — that is, the hypothesis that the selection differences in the natural aggregations involve not only differences in mean level, but differences in maturation rate.

This point can be illustrated in terms of the traditional quasi- experimental design problem of the effects of Latin on English vocabulary (Campbell 1963). In the hypothetical data of figure 6.10b, two alternative interpretations remain open. Latin may have had effect, for those taking Latin gained more than those not. But, on the other hand, those students taking Latin may have a greater annual rate of vocabulary growth that would manifest itself whether or not they took Latin. Extending this common design into two time series provides relevant evidence, as com- parison of the two alternative outcomes of figures 6.10c and 6.10d shows. Thus approaching quasi-experimental design from either improving the nonequivalent control-group design or from im- proving the interrupted time-series design, we arrive at the con- trol-series design. Figure 6.11 shows this for the Connecticut speeding crackdown, adding evidence from the fatality rates of neighboring states. Here the data are presented as population- based fatality rates so as to make the two series of comparable magnitude.

The control-series design of figure 6.11 shows that downward trends were available in the other states for 1955–56 as due to history and maturation, that is, due to shared secular trends, weather, automotive safety features, etc. But the data also show a general trend for Connecticut to rise relatively closer to the other states prior to 1955, and to steadily drop more rapidly than other states from 1956 on. Glass (1968) has used our monthly data for Connecticut and the control states to generate a monthly difference score, and this too shows a significant shift in trend in the Box and Tiao (1965) statistic. Impressed particularly by the 1957, 1958, and 1959 trend, we are willing to conclude that the crackdown had

FIGURE 6.10. Forms of Quasi-Experimental Analysis for the Effect of Specific Course Work, Including Control-Series Design

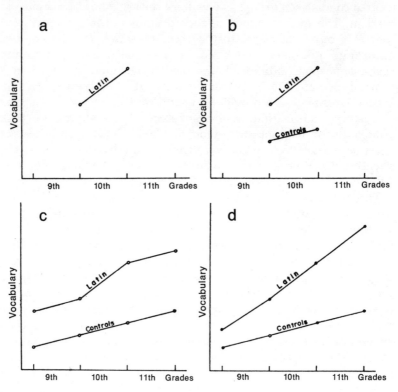

some effect, over and above the undeniable pseudo-effects of regression (Campbell and Ross 1968).

The advantages of the control-series design point to the advantages for social experimentation of a social system allowing subunit diversity. Our ability to estimate the effects of the speeding crackdown, Rose's (1952) and Stieber's (1949) ability to estimate the effects on strikes of compulsory arbitration laws, and Simon's (1966) ability to estimate the price elasticity of liquor were made possible because the changes were not being put into effect in all states simultaneously—because they were matters of state legislation rather than national. I do not want to appear to justify on these grounds the wasteful and unjust diversity of laws and enforcement practices from state to state. But I would strongly advocate that social engineers make use of this diversity while it remains available, and plan cooperatively their changes in ad-

FIGURE 6.11.   Control-Series Design Comparing Connecticut
Fatalities with Those of Four Comparable States

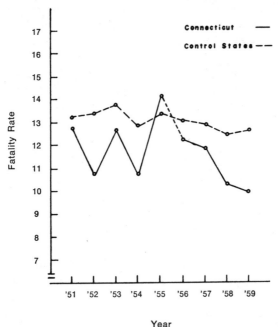

Year

ministrative policy and in record keeping so as to provide optimal
experimental inference. More important is the recommendation
that, for those aspects of social reform handled by the central
government, a purposeful diversity of implementation be envisaged
so that experimental and control groups be available for analysis.
Properly planned, these can approach true experiments, better
than the casual and ad hoc comparison groups now available. But
without such fundamental planning, uniform central control can
reduce the present possibilities of reality testing, that is, of
true social experimentation. In the same spirit, decentralization
of decision-making, both within large government and within
private monopolies, can provide a useful competition for both
efficiency and innovation, reflected in a multiplicity of indicators.

THE BRITISH BREATHALYSER CRACKDOWN

One further illustration of the interrupted time series and the
control series will be provided. The variety of illustrations so far

have each illustrated some methodological point, and have thus ended up as "bad examples." To provide a "good example," an instance which survives methodological critique as a valid illustration of a successful reform, data from the British Road Safety Act of 1967 are provided in figure 6.12 (Ross, Campbell, and Glass 1970).

The data on a weekly-hours basis are only available for a composite category of fatalities plus serious injuries, and figure 6.12 therefore uses this composite for all three bodies of data. The "weekend nights" figure comprises Friday and Saturday nights from 10:00 P.M. to 4:00 A.M. Here, as expected, the crackdown is most dramatically effective, producing initially more than a 40 per cent drop, leveling off at perhaps 30 per cent, although this involves dubious extrapolations in the absence of some control comparison to indicate what the trend over the years might have been without the crackdown. In this British case, no comparison state with comparable traffic conditions or drinking laws was available. But controls need not always be separate groups of persons; they may also be separate samples of times or stimulus materials (Campbell and Stanley 1966:43–47). A cigarette company may use the sales of its main competitor as a control comparison to evaluate a new advertising campaign. One should search around for the most nearly appropriate control comparison. For the Breathalyser crackdown, commuting hours when pubs had been long closed seemed ideal. (The "commuting hours" figures come from 7:00 A.M. to 10:00 A.M. and 4:00 P.M. to 5:00 P.M. Pubs are open for lunch from 12:00 to 2:00 or 2:30, and open again at 5:00 P.M.)

These commuting hours data convincingly show no effect, but are too unstable to help much with estimating the long-term effects. They show a different annual cycle than do the weekend nights or the over-all figures, and do not go back far enough to provide an adequate base for estimating this annual cycle with precision.

The use of a highly judgmental category such as "serious injuries" provides an opportunity for pseudo-effects due to a shift in the classifiers' standards. The over-all figures are available separately for fatalities, and these show a highly significant effect as strong as that found for the serious injury category or the composite shown in figure 6.12.

More details and the methodological problems are considered in our fuller presentation (Ross, Campbell, and Glass 1970). One rule for the use of this design needs reemphasizing. The inter-

FIGURE 6.12.   British Traffic Casualties (Fatalities plus Serious Injuries) Before and After the British Breathalyser Crackdown of October, 1967, Seasonally Adjusted. Modified from figure 1 of H. L. Ross, D. T. Campbell, and G. V. Glass, "Determining the Social Effects of a Legal Reform: The British 'Breathalyser' Crackdown of 1967" [*American Behavioral Scientist* 13 (1970):493– 509], on the basis of the revised and extended data reported in H. L. Ross, "Law, Science, and Accidents: The British Road Safety Act of 1967" [*Journal of Legal Studies,* 2 (1973):1–78], and L. J. McCain, "Technical Appendix: Preparation and Analysis of Data in an Interrupted Time Series Quasi-Experiment," in *Law, Science, and Accidents: The British Road Safety Act of 1967,* ed. H. L. Ross [Chicago: American Bar Foundation, forthcoming.]

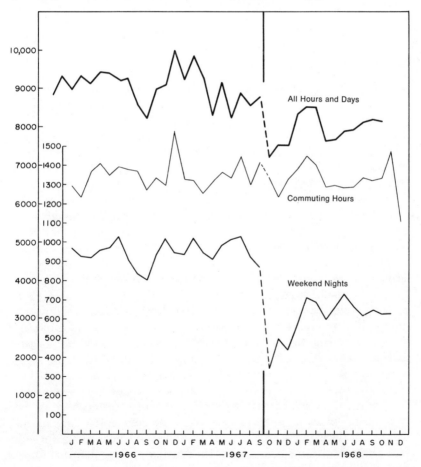

rupted time series can only provide clear evidence of effect where the reform is introduced with a vigorous abruptness. A gradually introduced reform has little chance of being distinguished from shifts in secular trends or from the cumulative effect of the many other influences impinging during a prolonged period of introduction. In the Breathalyser crackdown, an intense publicity campaign naming the specific starting date preceded the actual crackdown. Although the impact seems primarily due to publicity and fear rather than to an actual increase of arrests, an abrupt initiation date was achieved. Had the enforcement effort changed at the moment the act had passed, with public awareness being built up by subsequent publicity, the resulting data series would have been essentially uninterpretable.

REGRESSION DISCONTINUITY DESIGN

We shift now to social ameliorations that are in short supply, and that therefore cannot be given to all individuals. Such scarcity is inevitable under many circumstances, and can make possible an evaluation of effects that would otherwise be impossible. Consider the heroic Salk poliomyelitis vaccine trials in which some children were given the vaccine while others were given an inert saline placebo injection – and in which many more of these placebo controls would die than would have if they had been given the vaccine. Creation of these placebo controls would have been morally, psychologically, and socially impossible had there been enough vaccine for all. As it was, due to the scarcity, most children that year had to go without the vaccine anyway. The creation of experimental and control groups was the highly moral allocation of that scarcity so as to enable us to learn the true efficacy of the supposed good. The usual medical practice of introducing new cures on a so-called trial basis in general medical practice makes evaluation impossible by confounding prior status with treatment, that is, giving the drug to the most needy or most hopeless. It has the further social bias of giving the supposed benefit to those most assiduous in keeping their medical needs in the attention of the medical profession, that is, the upper and upper-middle classes. The political stance furthering social experimentation here is the recognition of randomization as the most democratic and moral means of allocating scarce resources (and scarce hazardous duties), plus the moral imperative to further utilize the randomization so that society may indeed learn the true value of the supposed boon.

This is the ideology that makes possible "true experiments" in a large class of social reforms.

But if randomization is not politically feasible or morally justifiable in a given setting, there is a powerful quasi-experimental design available that allows the scarce good to be given to the most needy or the most deserving. This is the regression discontinuity design. All it requires is strict and orderly attention to the priority dimension. The design originated through an advocacy of a tie-breaking experiment to measure the effects of receiving a fellowship (Thistlethwaite and Campbell 1960), and it seems easiest to explain it in that light. Consider, as in figure 6.13, pre-

FIGURE 6.13. Tie-Breaking Experiment and Regression Discontinuity Analysis

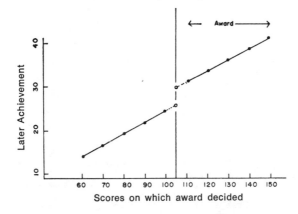

award ability-and-merit dimension, which would have some relation to later success in life (finishing college, earnings ten years later, etc.). Those higher on the premeasure are most deserving and receive the award. They do better in later life, but does the award have an effect? It is normally impossible to say because they would have done better in later life anyway. Full randomization of the award was impossible given the stated intention to reward merit and ability. But it might be possible to take a narrow band of ability at the cutting point, to regard all of these persons as tied, and to assign half of them to awards, half to no awards, by means of a tie-breaking randomization.

The tie-breaking rationale is still worth doing, but in considering that design it became obvious that, if the regression of premeasure on later effects were reasonably orderly, one should be able to extrapolate to the results of the tie-breaking experiment by

FIGURE 6.14.   Illustrative Outcomes of Regression Discontinuity
Analyses

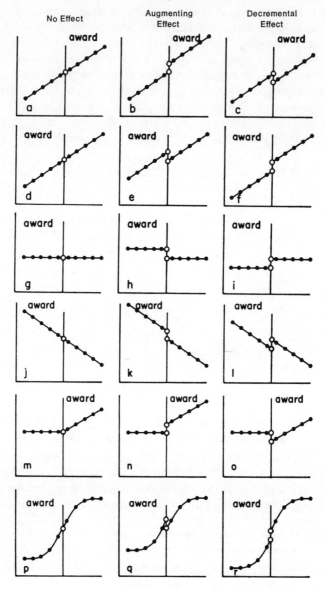

plotting the regression of posttest on pretest separately for those in the award and nonaward regions. If there is no significant difference for these at the decision-point intercept, then the tie-breaking experiment should show no difference. In cases where the tie breakers would show an effect, there should be an abrupt discontinuity in the regression line. Such a discontinuity cannot be explained away by the normal regression of the posttest on pretest, for this normal regression, as extensively sampled within the nonaward area and within the award area, provides no such expectations.

Figure 6.13 presents, in terms of column means, an instance in which higher pretest scores would have led to higher posttest scores even without the treatment, and in which there is in addition a substantial treatment effect. Figure 6.14 shows a series of paired outcomes, those on the left to be interpreted as no effect, those in the center and on the right as effect. Note some particular cases. In instances of granting opportunity on the basis of merit, like 6.14a and b (and figure 6.13), neglect of the background regression of pretest on posttest leads to optimistic pseudo-effects: in figure 6.14a, those receiving the award do do better in later life, though not really because of the award. But in social ameliorative efforts, the setting is more apt to be like figure 6.14d and e, where neglect of the background regression is apt to make the program look deleterious if there is no effect, or ineffective if there is a real effect.

The design will of course work just as well or better if the award dimension and the decision base, the pretest measure, are unrelated to the posttest dimension, if it is irrelevant or unfair, as instanced in figure 6.14g, h, and i. In such cases the decision base is the functional equivalent of randomization. Negative background relationships are obviously possible, as in figure 6.14j, k, and l. In figure 6.14 m, n, and o are included to emphasize that it is a jump in intercept at the cutting point that shows effect, and that differences in slope without differences at the cutting point are not acceptable as evidences of effect. This becomes more obvious if we remember that in cases like 6.14m, a tie-breaking randomization experiment would have shown no difference. Curvilinear background relationships, as in figure 6.14p, q, and r, will provide added obstacles to clear inference in many instances, where sampling error could make figure 6.14p look like 6.14b.

As further illustration, figure 6.15 provides computer-simulated data, showing individual observations and fitted regression lines, in a fuller version of the no-effect outcome of figure 6.14a. Figure

FIGURE 6.15.   Regression Discontinuity Design: No Effect

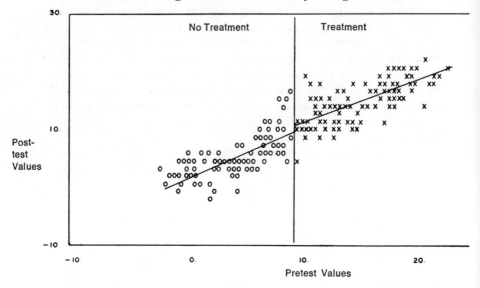

6.16 shows an outcome with effect. These have been generated by assigning to each individual a weighted normal random number as a "true score," to which is added a weighted independent "error" to generate the "pretest." The "true score" plus another independent "error" produces the "posttest" in no-effect cases such as figure 6.15. In treatment-effect simulations, as in figure 6.16, there are added into the posttest "effects points" for all "treated" cases, that is, those above the cutting point on the pretest score.

This design could be used in a number of settings. Consider Job Training Corps applicants, in larger number than the program can accommodate, with eligibility determined by need. The setting would be as in figure 6.14d and e. The base-line decision dimension could be per capita family income, with those at below the cutoff getting training. The outcome dimension could be the amount of withholding tax withheld two years later, or the percentage drawing unemployment insurance, these follow-up figures being provided from the National Data Bank in response to categorized social security numbers fed in, without individual anonymity being breached, without any real invasion of privacy—by the technique of mutually insulated data banks. While the plotted points could be named, there is no need that they be named. In a classic field experiment on tax compliance, Richard Schwartz and the Bureau of Internal Revenue have managed to put together

FIGURE 6.16.   Regression Discontinuity Design: Genuine Effect

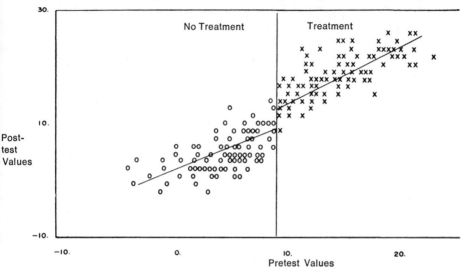

Pretest Values

sets of personally identified interviews and tax-return data so that statistical analyses such as these could be done, without the separate custodians of either interview or tax returns learning the corresponding data for specific persons (Schwartz and Orleans 1967; Schwartz and Skolnick 1963; Levenson and McDill 1966).

Applied to the Job Corps illustration, it would work as follows: Separate lists of Job Corps applicants (with social security numbers) would be prepared for every class interval on per capita family income. To each of these lists an alphabetical designation would be assigned at random. (Thus the $10.00 per week list might be labeled *M;* $11.00, *C;* $12.00, *Z;* $13.00, *Q;* $14.00, *N;* etc.) These lists would be sent to Internal Revenue, without the Internal Revenue personnel being able to learn anything interpretable about the applicants' traineeship status or family income. The Internal Revenue statisticians would locate the withholding tax collected for each person on each list, but would not return the data in that form. Instead, for each list, only the withholding tax amounts would be listed, and these in a newly randomized order. These would be returned to Job Corps research, who could use them to plot a graph similar to figures 6.10, 6.11, and 6.12, and do the appropriate statistical analyses by retranslating the alphabetical symbols into meaningful base-line values. But within any list, they would be unable to learn which value belonged to which person. (To insure this effective anonymity, it could be specified that no lists shorter than 100 persons be used, the base-line intervals being expanded if necessary to achieve this.) Manniche and Hayes (1957) have spelled out how a broker can be used in a two-staged matching of doubly coded data. Kaysen (1967), Sawyer and Schechter (1968), and Boruch (1971) have wise discussions of the more general problem.

What is required of the administrator of a scarce ameliorative commodity to use this design? Most essential is a sharp cutoff point on a decision-criterion dimension, on which several other qualitatively similar analytic cutoffs can be made both above and below the award cut. Let me explain this better by explaining why National Merit scholarship administrators were unable to use the design for their actual fellowship decision (although it has been used for their Certificate of Merit). In their operation, diverse committees make small numbers of award decisions by considering a group of candidates and then picking from them the N best to which to award the N fellowships allocated them. This provides one cutting point on an unspecified pooled decision base, but fails to provide analogous potential cutting points above and below. What could be done is for each committee to collectively rank its group of 20 or so candidates. The top N would then receive the award. Pooling cases across committees, cases could be classified according to number of ranks above and below the cutting point, these other ranks being analogous to the award-nonaward cutting point as far as regression onto posttreatment measures was concerned. Such group ranking would be costly in terms of committee time. An equally good procedure, if committees agreed, would be to have each member, after full discussion and freedom to revise, give each candidate a grade—A+, A, A−, B+, B, etc.—and to award the fellowships to the N candidates averaging best on these ratings, with no revisions allowed after the averaging process. These ranking or rating units, even if not comparable from committee to committee in range of talent, in number of persons ranked, or in cutting point, could be pooled without bias as far as a regression discontinuity is concerned, for that range of units above and below the cutting point in which all committees were represented.

It is the dimensionality and sharpness of the decision criterion that is at issue, not its components or validity. The ratings could be based upon nepotism, whimsey, and superstition and still serve. As has been stated, if the decision criterion is utterly invalid we approach the pure randomness of a true experiment. Thus the weakness of subjective committee decisions is not their subjectivity, but the fact that they provide only the one cutting point on their net subjective dimension. Even in the form of average ratings the recommended procedures probably represent some slight increase in committee work load. But this could be justified to the decision committees by the fact that through refusals, etc., it cannot be known at the time of the committee meeting the exact number to whom the fellowship can be offered. Other costs at the planning time are likewise minimal. The primary additional burden is in keeping as good records on the nonawardees as on the awardees. Thus at a low cost, an experimental administrator can lay the groundwork for later scientific follow-ups, the budgets for which need not yet be in sight.

Our present situation is more apt to be one where our pretreatment measures, aptitude measures, reference ratings, etc., can be combined via multiple correlation into an index that correlates highly but not perfectly with the award decision. For this dimension there is a fuzzy cutoff point. Can the design be used in this case? Probably not. Figure 6.17 shows the pseudo-effect possible if the award decision contributes any valid variance to the quantified pretest evidence, as it usually will. The award regression rides above the nonaward regression just because of that valid variance in this simulated case, there being no true award effect at all. (In simulating this case, the award decision has been

FIGURE 6.17.   Regression Discontinuity Design: Fuzzy Cutting
Point, Pseudo Treatment Effect Only

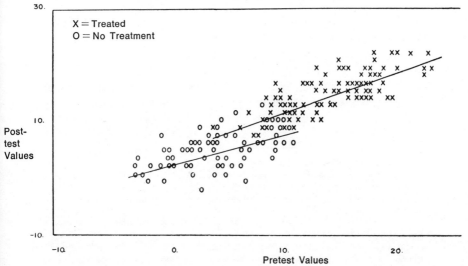

FIGURE 6.18. Regression Discontinuity Design: Fuzzy Cutting
Point, with Real Treatment Plus Pseudo Treatment Effects

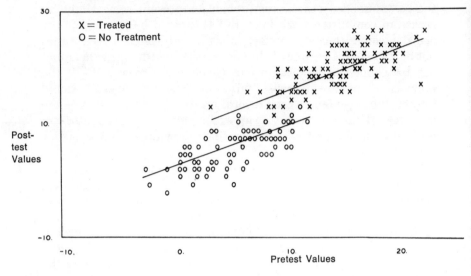

based upon a composite of true score plus an independent award
error.) Figure 6.18 shows a fuzzy cutting point plus a genuine
award effect.[5] The recommendation to the administrator is clear:
aim for a sharp cutting point on a quantified decision criterion. If
there are complex rules for eligibility, only one of which is quantified, seek out for follow-up that subset of persons for whom the
quantitative dimension was determinate. If political patronage
necessitates some decision inconsistent with a sharp cutoff, record
these cases under the heading "qualitative decision rule" and keep
them out of your experimental analysis.

Almost all of our ameliorative programs designed for the disadvantaged could be studied via this design, as could some major
governmental actions affecting the lives of citizens in ways we do
not think of as experimental. For example, for a considerable
period, quantitative test scores have been used to determine whom

5. There are some subtle statistical clues that might distinguish these two instances if one had enough cases. There should be increased pooled column variance
in the mixed columns for a true effects case. If the data are arbitrarily treated as
though there had been a sharp cutting point located in the middle of the overlap
area, then there should be no discontinuity in the no-effect case, and some discontinuity in the case of a real effect, albeit an underestimated discontinuity, since
there are untreated cases above the cutting point and treated ones below, dampening the apparent effect. The degree of such dampening should be estimable, and
correctable, perhaps by iterative procedures. But these are hopes for the future.

to call up for military service, or to reject as unfit at the lower ability range. If these cutting points, test scores, names, and social security numbers have been recorded for a number of steps both above and below the cutting point, we could make elegant studies of the effect of military service on later withholding taxes, mortality, number of dependents, etc.

This illustration points to one of the threats to external validity of this design, or of the tie-breaking experiment. The effect of the treatment has only been studied for that narrow range of talent near the cutting point, and generalization of the effects of military service, for example, from this low ability level to the careers of the most able would be hazardous in the extreme. But in the draft laws and the requirements of the military services there may be other sharp cutting points on a quantitative criterion that could also be used. For example, those over 6 feet 6 inches are excluded from service. Imagine a five-year-later follow-up of draftees grouped by inch in the 6 feet 1 inch to 6 feet 5 inches range, and a group of their counterparts who would have been drafted except for their heights, 6 feet 6 inches to 6 feet 10 inches. (The fact that the other grounds of deferment might not have been examined by the draft board would be a problem here, but probably not insurmountable.) That we should not expect height in this range to have any relation to later-life variables is not at all a weakness of this design, and if we have indeed a subpopulation for which there is a sharp numerical cutting point, an internally valid measure of effects would result. Deferment under the present system is an unquantified committee decision. But just as the sense of justice of United States soldiers was quantified through paired comparisons of cases into an acceptable Demobilization Points system at the end of World War II (Guttman 1946; Stouffer 1949), so a quantified composite index of deferment priority could be achieved and applied as uniform justice across the nation, providing another numerical cutting point.

In addition to the National Data Bank type of indicators, there will be occasions in which new data collections, perhaps by interview or questionnaire, are needed. For these there is the special problem of uneven cooperation that would be classified as instrumentation error. In our traditional mode of thinking, completeness of description is valued more highly than comparability. Thus if, in a fellowship study, a follow-up mailed out from the fellowship office would bring a higher return from past winners, this might seem desirable even if the nonawardees' rate of response was much lower. From the point of view of quasi-experimentation, however, it would be better to use an independent survey agency and a disguised purpose, achieving equally low response rates from both awardees and nonawardees, and avoiding a regression discontinuity in cooperation rate that might be misinterpreted as a discontinuity in more important effects.

RANDOMIZED CONTROL GROUP EXPERIMENTS

Experiments with randomization tend to be limited to the laboratory and agricultural experiment station. But this certainly need not be so. The randomization unit may be persons, families, precincts, or larger administrative units. For statistical purposes the randomization units should be numerous, and hence ideally small. But for reasons of external validity, including reactive arrangements, the randomization units should be selected on the basis of the units of administrative access. Where policies are administered through individual client contacts, randomization at the person level may often be inconspicuously achieved, with the clients unaware that different ones are getting different treatments. But for most social reforms, larger administrative units will be involved, such as classrooms, schools, cities, counties, or states. We need to develop the political postures and ideologies that make randomization at these levels possible.

"Pilot project" is a useful term already in our political vocabulary. It designates a trial program that, if it works, will be spread to other areas. By modifying actual practice in this regard, without going outside of the popular understanding of the term, a valuable experimental ideology could be developed. How are areas selected for pilot projects? If the public worries about this, it probably assumes a lobbying process in which the greater needs of some areas are only one consideration, political power and expediency being others. Without violating the public tolerance or intent, one could probably devise a system in which the usual lobbying decided upon the areas eligible for a formal public lottery that would make final choices between matched pairs. Such decision procedures as the drawing of lots have had a justly esteemed position since time immemorial (e.g., Aubert 1959). At the present time, record keeping for pilot projects tends to be limited to the experimental group only. In the experimental ideology, comparable data would be collected on designated controls. (There are, of course, exceptions, as in the heroic Public Health Service fluoridation experiments, in which the teeth of Oak Park children were examined year after year as controls for the Evanston experiments [Blayney and Hill 1967].)

Another general political stance making possible experimental social amelioration is that of *staged innovation*. Even though by intent a new reform is to be put into effect in all units, the logistics of the situation usually dictate that simultaneous introduction is not possible. What results is a haphazard sequence of convenience. Under the program of staged innovation, the introduction of the program would be deliberately spread out, and those units selected to be first and last would be randomly assigned (perhaps randomization from matched pairs), so that during the transition period the first recipients could be analyzed as experimental units, the last recipients as controls. A third ideology making possible true experiments has already been discussed: randomization as the democratic means of allocating scarce resources.

This article will not give true experimentation equal space with quasi-experimentation only because excellent discussions of, and statistical consultation on, true experimentation are readily available. True experiments should almost always be preferred to quasi-experiments where both are available. Only occasionally are the threats to external validity so much greater for the true experiment that one would prefer a quasi-experiment. The uneven allocation of space here should not be read as indicating otherwise.

MORE ADVICE FOR TRAPPED ADMINISTRATORS

But the competition is not really between the fairly interpretable quasi-experiments here reviewed and "true" experiments. Both stand together as rare excellencies in contrast with a morass of obfuscation and self-deception. Both to emphasize this contrast, and again as guidelines for the benefit of those trapped administrators whose political predicament will not allow the risk of failure, some of these alternatives should be mentioned.

GRATEFUL TESTIMONIALS

Human courtesy and gratitude being what they are, the most dependable means of assuring a favorable evaluation is to use voluntary testimonials from those who have had the treatment. If the spontaneously produced testimonials are in short supply, these should be solicited from the recipients with whom the program is still in contact. The rosy glow resulting is analogous to the professor's impression of his teaching success when it is based solely upon the comments of those students who come up and talk with him after class. In many programs, as in psychotherapy, the recipient, as well as the agency, has devoted much time and effort to the program and it is dissonance-reducing for himself, as well as common courtesy to his therapist, to report improvement. These grateful testimonials can come in the language of letters and conversation, or be framed as answers to multiple-item "tests" in which a recurrent theme of "I am sick," "I am well," "I am happy," "I am sad" recurs. Probably the testimonials will be more favorable: (1) the more the evaluative meaning of the response measure is clear to the recipient – it is completely clear in most personality, adjustment, morale, and attitude tests; (2) the more directly the recipient is identified by name with his answer; (3) the more the recipient gives the answer directly to the therapist or agent of reform; (4) the more the agent will continue to be influential in the recipient's life in the future; (5) the more the answers deal with feelings and evaluations rather than with verifiable facts; and (6) the more the recipients participating in the evaluation are a small and self-selected or agent-selected subset of all recipients. Properly designed, the grateful testimonial method can involve pretests as well as posttests, and randomized control groups as well as experimentals, for there are usually no placebo treatments, and the recipients know when they have had the boon.

CONFOUNDING SELECTION AND TREATMENT

Another dependable tactic bound to give favorable outcomes is to confound selection and treatment, so that in the published comparison those receiving the treatment are also the more able and well placed. The often-cited evidence of the dollar value of a college education is of this nature – all careful studies show that most of the effect, and of the superior effect of superior colleges, is explainable in terms of superior talents and family connections, rather than in terms of what is learned or even the prestige of the degree. Matching techniques and statistical partialings generally undermatch and do not fully control for the selection differences – they introduce regression artifacts confusable as treatment effects.

There are two types of situations that must be distinguished. First, there are those treatments that are given to the most promising, treatments like a college education which are regularly given to those who need it least. For these, the later concomitants of the grounds of selection operate in the same direction as the treatment: those most likely to achieve anyway get into the college most likely to produce later achievement. For these settings, the trapped administrator should use the pooled mean of all those treated, comparing it with the mean of all untreated, although in this setting almost any comparison an administrator might hit upon would be biased in his favor.

At the other end of the talent continuum are those remedial treatments given to those who need it most. Here the later concomitants of the grounds of selection are poorer success. In the Job Training Corps example, casual comparisons of the later unemployment rate of those who received the training with those who did not are in general biased against showing an advantage to the training. This seems to have been the case in the major Head Start evaluation (Campbell and Erlebacher 1970).

Here the trapped administrator must be careful to seek out those few special comparisons biasing selection in his favor. For training programs such as Operation Head Start and tutoring programs, a useful solution is to compare the later success of those who completed the training program with those who were invited but never showed plus those who came a few times and dropped out. By regarding only those who complete the program as "trained" and using the others as controls, one is selecting for conscientiousness, stable and supporting family backgrounds, enjoyment of the training activity, ability, determination to get ahead in the world — all factors promising well for future achievement even if the remedial program is valueless. To apply this tactic effectively in the Job Training Corps, one might have to eliminate from the so-called control group all those who quit the training program because they had found a job — but this would seem a reasonable practice and would not blemish the reception of a glowing progress report.

These are but two more samples of well-tried modes of analysis for the trapped administrator who cannot afford an honest evaluation of the social reform he directs. They remind us again that we must help create a political climate that demands more rigorous and less self-deceptive reality testing. We must provide political stances that permit true experiments, or good quasi-experiments. Of the several suggestions toward this end that are contained in this article, the most important is probably the initial theme: Administrators and parties must advocate the importance of the problem rather than the importance of the answer. They must advocate experimental sequences of reforms, rather than one certain cureall, advocating reform A with alternative B available to try next should an honest evaluation of A prove it worthless or harmful.

MULTIPLE REPLICATION IN ENACTMENT

Too many social scientists expect single experiments to settle issues once and for all. This may be a mistaken generalization from the history of great crucial experiments in physics and chemistry. In actuality the significant experiments in the physical sciences are replicated thousands of times, not only in deliberate replication efforts, but also as inevitable incidentals in successive experimentation and in utilizations of those many measurement devices (such as the galvanometer) that in their own operation embody the principles of classic experiments. Because we social scientists have less ability to achieve "experimental isolation," because we have good reason to expect our treatment effects to interact significantly with a wide variety of social factors, many of which we have not yet mapped, we have much greater needs for replication experiments than do the physical sciences.

The implications are clear. We should not only do hardheaded reality testing in the initial pilot testing and choosing of which reform to make general law; once it has been decided that the reform is to be adopted as standard practice in all administrative units, we should experimentally evaluate it in each of its implementations (Campbell 1967).

CONCLUSIONS

Trapped administrators have so committed themselves in advance to the efficacy of the reform that they cannot afford honest evaluation. For them, favorably biased analyses are recommended, including capitalizing on regression, grateful testimonials, and confounding selection and treatment. *Experimental administrators* have justified the reform on the basis of the importance of the problem, not the certainty of their answer, and are committed to going on to other potential solutions if the one first tried fails. They are therefore not threatened by a hardheaded analysis of the reform. For such, proper administrative decisions can lay the base for useful experimental or quasi-experimental analyses. Through the ideology of allocating scarce resources by lottery, through the use of staged innovation, and through the pilot project, true experiments with randomly assigned control groups can be achieved. If the reform must be introduced across the board, the interrupted time-series design is available. If there are similar units under independent administration, a control-series design adds strength. If a scarce boon must be given to the most needy or to the most deserving, quantifying this need or merit makes possible the regression discontinuity analysis.

7

ORGANIZATIONAL MEMBERSHIP AND ATTITUDE CHANGE

RICHARD D. SHINGLES

Students of political behavior and contemporary political systems have long been interested in the role of secondary associations in influencing political orientations. One of the principal generalizations stemming from work in this area is that voluntary membership in autonomous organizations leads to a frame of mind and a set of behavioral patterns which are essential to the democratic process. The nature and quality of the rationales provided for this conclusion vary, depending upon the author. A common focus, however, is on a causal relationship in which *involvement in secondary associations (A)* is said to lead to *increased political involvement (C)* of a type which is consistent with, and beneficial to, the democratic process. The occurrence of the relationship is typically accredited to *the development of specific attitudes (B)*, such as a sense of political efficacy or a belief in regime legitimacy, which stem from involvement in such associations and in turn serve to encourage greater levels of democratic participation. The over-all relationship may be symbolized as follows:

$$A \xrightarrow{(+)} B \xrightarrow{(+)} C, \quad \text{or} \quad \text{Structure} \xrightarrow{(+)} \text{Attitudes} \xrightarrow{(+)} \text{Behavior}$$

Individual variants of this basic theme have provided a useful and interesting explanation of both political behavior and the nature of political systems. Nevertheless, they lack the empirical support required for their acceptance as valid interpretations of reality. This is primarily because the methodology which has traditionally been used to substantiate them is unable to provide data which can differentiate between this theme and other equally plausible theses as to the relationship between organizational involvement, psychological orientations, and political activity. It is the purpose of this paper to review these limitations, and to offer additional evidence collected from a quasi-experiment which is designed to circumvent problems of multiple interpretations. My attention will center on two variants of the $A \xrightarrow{(+)} B \xrightarrow{(+)} C$ theme: the civic culture thesis of Almond and Verba (1963) and the mass society thesis as represented in the works of Seeman (1966), Neal and Seeman (1964), and Lipset, Trow, and Coleman (1956).

The discussion is divided into four parts. The first section presents the basic tenets put forth by these authors. A second section is concerned with the weakness in the evidence offered in support of the theory, as well as with the plausibility of some alternative formulations. A quasi-experimental research design believed appropriate for explicitly evaluating some of these alternatives is outlined in the third section. Finally, results of research based on this design are presented. The analysis provided in this last section will focus solely on the A–B linkage. For a specific case, a St. Paul community action agency, it attempts to answer the question "Does involvement affect attitudes or do attitudes predispose one to involvement?"

THE MASS SOCIETY AND CIVIC CULTURE THESES

Lipset, Trow, and Coleman provide a unique test of the mass society thesis in their application of it to the internal politics of large-scale organizations. In doing so they also illustrate an alternative to Michels' "iron law of oligarchy" (1949). Large-scale organizations are commonly characterized by oligarchies in which the incumbent leadership permanently maintains itself in power through a monopoly of status, skill, information, and its position in the organizational bureaucracy. Such advantages make control over the organization by the rank and file practically impossible, thereby contributing to considerable mass apathy regarding internal organizational politics. Mass society theorists contend, however, that the tendency toward oligarchy or dictatorship either in societies at large or in specific organizations can be overcome by the presence of numerous, autonomous, voluntary organizations. The presence of such associations provides an independent base of power necessary for sustaining an effective opposition to incumbent elites. In addition, they maintain an informal communication system, a training ground for the development of organizational and political skills, and a common basis for social interaction and collective power. It is assumed that such structural characteristics help to create and sustain an interest in and awareness of politics and a sense of confidence that political action can be both meaningful and rewarding.

Lipset, Trow, and Coleman base these conclusions primarily on a study of the International Typographical Union. They explain the existence of a viable democracy within the organization (characterized by a two-party system and active rank-and-file participation in union politics) as due to a network of voluntary organizations (such as printers' social clubs, lodges, sports clubs, and veterans' groups) which have developed without any formal connection with the union or its political apparatus. The authors find a close association between participation in the informal printers' community and involvement in union politics ($A \overset{(+)}{\text{——}} C$) and interpret this to mean that the former contributes to the latter ($A \overset{(+)}{\longrightarrow} C$). The question of intervening attitudinal variables, however, is left largely unexplored.

The contribution to this discussion of both Melvin Seeman and Almond and Verba is their efforts to develop the rationale linking attitudes with organizational membership and political participation and to provide empirical grounds for supporting their thesis. They base the $A \overset{(+)}{\longrightarrow} C$ contention on individual perceptions of personal competence. Although they employ different terms — "feelings of power or powerlessness" in the former and "sense of political competence" in the latter case — they are essentially discussing the same concept, more commonly termed "political efficacy" (Campbell et al. 1960). This is the feeling on the part of the individual that as a citizen he has some ability to affect political decisions, that the system is amenable to political influence, that officials are or can be made responsible to individual appeals. Sense of powerlessness is the inverse of competence or efficacy. Throughout the remainder of this paper these terms — competence and efficacy — will be used interchangeably with the word *alienation*, the latter being taken to mean principally a sense of inefficacy or powerlessness.

According to Seeman, sense of personal efficacy is the key variable in accounting for the effect of organizational membership (or lack of it) on political involvement. In modernizing and advanced technological societies, the destruction of traditional communities, the breakdown of binding social ties, and the growth of bureaucracy lead the increasingly more isolated individual to become convinced of his own personal powerlessness (i.e., "the expectancy or probability held by the individual that his own behavior cannot determine the occurrence of the outcome he seeks"). As a consequence, he sees no utility in, and becomes apathetic about, control relevant learning (such as the acquisition of knowledge about political issues and events which directly affect him) and control relevant behavior (such as political activity in pursuing personal or public goals). The remedy to such a situation, according to Seeman, lies in the development of, and membership in, secondary associations which mediate between the individual and the society. Those individuals who become members of such associations will experience a greater feeling of personal effectiveness — of which political efficacy is one dimension[1] — and consequently will be more likely to become politically involved. The major inference is that, once a member of such groups, the individual will in fact be more powerful as part of a collective, and that he will sense this power and identify with it.

Almond and Verba take a similar approach — one which adds an important element to the theory. In their thesis, membership in an association does not by itself enhance one's sense of efficacy. The important factor is the degree of participation in formulating group decisions. Those individuals who are, or who feel they are, helping to make such decisions will acquire a sense of personal competence which is then transferred to the political arena. The basis of their generalization is the principle of stimulus generalization (Hull 1950) in which a reaction (heightened sense of efficacy) to one situation (the ability to participate in an association) is generalized by the individual from one context (the organization) to another (the political system) (Almond and Verba 1963:136 ff., 244–45, 346). The importance of political efficacy for a stable democratic process (i.e., creating an interested, well-informed, and active citizenry) is said not only to lie in the incentive it provides to actively participate politically, but also in the fact that a heightened sense of political competence leads to greater satisfaction with the system as a whole. The rationale is that legitimacy stems from one's belief that he has some control over systemwide decisions (Almond and Verba 1963:187–202).

1. Supporting this contention, Campbell et al. (1960:515–19) have demonstrated a high positive association between a general sense of personal effectiveness and a sense of political efficacy.

AVAILABLE EVIDENCE AND
ALTERNATIVE HYPOTHESES

The principal empirical drawback of these studies is that, with one exception, the type of empirical support made available by them is based solely on data obtained from interviews taken at one point in time. The limitations of cross-sectional data for making causal inferences have been presented by Roos in chapter 2. Lacking temporal priority, it is difficult to proceed from the simple demonstration of association to causation. Lipset, Trow, and Coleman base their conclusions on correlations revealing a high positive association between membership in voluntary associations in the ITU (A) and political involvement (C). They then infer that A caused C and explain the connection in terms of a hypothetical, but nonmeasured, B. Almond and Verba and Seeman go further, empirically demonstrating the importance of B (efficacy). They demonstrate that A is positively associated with B and that high levels of B are in turn associated with high levels of C. Once again, however, they must make a questionable inference that it is A that causes increases in B and that high levels of B increase one's propensity for C. In short, the causality is tenuously inferred in each case. Given the great difficulty of creating convincing empirical proofs of causality, such inferences are normally quite acceptable in the social sciences *if* the inferences are well grounded in theory. In such cases, the theoretical argument compensates for insufficiencies in the empirical analysis. However, in the case of these studies, the theoretical argument is also insufficient, principally because the competing interpretations are equally convincing. This leads one to believe that the direction the association takes is other than the one hypothesized, and thereby creates the need for a more rigorous empirical proof.

For example, it is quite plausible that instead of the causal relationship being $A \overset{(+)}{\longrightarrow} B \overset{(+)}{\longrightarrow} C$, it is $A \overset{(+)}{\longleftarrow} B \overset{(+)}{\longleftarrow} C$. That is, it is entirely possible that the experience of participating in the political process leads one to make new acquaintances, expands one's horizons, and ultimately leads to increased participation in other forms of social activity. It is equally plausible that participation in the political process leads to an enhanced sense of personal and political efficacy (in the same manner that Almond and Verba suggested participation in any decision-making process of secondary associations increases civic competence), and that this rise in one's self-confidence further leads to bolder experimentation with other forms of social involvement.[2]

Another convincing hypothesis is that A is actually irrelevant to C, that the relationship found between them is spurious, and that both are in fact a consequence of the third variable, B. In other words, it may be that certain types of people have a greater propensity to become socially involved than other types, and that involvement may take either political or nonpolitical forms or both depending upon the interest or talents of the individual. In this case, the path of causality would be as follows:

$$A^{(+)} \quad {}^{(+)}C$$
$$\nwarrow \quad \nearrow$$
$$B$$

The type of orientation that would lead to both A and C may be any one of several variables, including high feelings of personal competence, civic duty, other-directedness or gregariousness. If indeed the principal variable is efficacy, as it would seem given its association with involvement found by the studies cited previously, then these feelings may in turn be largely a product of still some other variable, such as level of educational attainment.

EFFORTS TO OVERCOME THE PROBLEM OF CAUSAL INFERENCE

Mass society–civic culture theorists vary considerably in their attempts to deal with the problem of causal inference. Lipset, Trow, and Coleman do employ a panel which normally might resolve this type of problem by allowing one to assume causality on the basis of temporal order. Yet because they are either not aware of, or not concerned with, the alternative hypotheses, and because they fail to employ an experimental design, their use of the panel fails to provide sufficient clarification. They demonstrate that over a six-month period (between the initial personal interview and a fol-

2. Similar problems of interpretation exist with the various forms of supportive data the authors employ. Finding positive associations between ideological sensitivity (liberal-conservative conceptualization) and political involvement in the ITU, Lipset, Trow, and Coleman interpret this to mean that the former causes the latter (1956:92–94). Yet Campbell et al. (1960:251) and McClosky (1964:374–75) both interpret the same relationship to mean that ideological sensitivity is a consequence of active political involvement. Similarly, Almond and Verba's contention that people who feel efficacious will believe the system to be legitimate makes as much or more sense if we reverse it to say that people who feel the system to be legitimate will therefore feel a higher sense of efficacy. In all probability, there is a mutual influence between these variables which existing methods of research have been unable to separate analytically.

low-up mail questionnaire) those individuals who had formerly been shown to be active club members reveal a greater increase in interest, knowledge, and activity in union politics than those who were not active in the printers' community (Lipset, Trow, and Coleman 1956:84–86). Unfortunately, during this same six-month period there were both national and union election campaigns going on, and all union members—whether members of the printers' community or not—show an increased involvement in both national and union politics.

Trends of this kind, in which election campaigns progressively boost the political interests and activity of voters, are typical (Berelson, Lazarsfeld, and McPhee 1954:33–34). In the case of the ITU, participants in the printers' secondary associations do reveal a relatively greater increase than nonparticipants, but using the alternative hypotheses, this could be interpreted to mean that the same people who have a high propensity to become socially active (whether the activity is political or nonpolitical) are also more sensitive to the stimulus of the campaign. It is known, for example, that under normal conditions the same active and politically sensitive individuals react to the stimulus of election campaigns whereas the apathetic and uninterested are only reached where short-term political forces are unusually intense (Angus Campbell 1966:40–62). Members of the printers' community may tend to show a greater increase in political involvement not because of their membership in the community, but because of the very reasons they become part of the community—the fact that they are activist types: outgoing, socially involved people who are interested in, and sensitive to, the world about them.

Perhaps the authors' complacency stems from their finding that the younger men in the ITU are more likely to be active in the social community while the older men are more prone to enter union politics. Since political participation generally does tend to be highest among middle-aged groups and lowest among young adults (Campbell et al. 1960:493–98), this could mean, as Lipset, Trow, and Coleman suggest, that younger men first participate in the printers' community and, as a direct consequence, later become involved in politics. On the other hand, it could simply indicate that certain types of people are more apt to be socially active; but whereas the younger men are attracted to social and sports activities, they become more interested in union politics as they grow older and develop more interest in the union and their job as a permanent livelihood.

The limitations of their data also prohibited Almond and Verba from considering the problem of the alternative hypotheses. In interviews, they asked the respondents to *recall* their ability to participate in their family as a child, in school, and in their place of work. Finding a positive association between recollection of this ability and the respondent's existing sense of civic competence, the authors assumed that the latter state of mind is caused by the prior experiences. The inference is based on temporal priority, but the assumption of temporal priority is questionable in that it is dependent on the accuracy of recall. A question occurs as to what extent the findings reflect the fact that degree of participation affects efficacy and to what extent it reflects a tendency on the part of generally efficacious, optimistic people to see the world in these terms and to recall a brighter past than those who are alienated, hostile, or forlorn.

Of all the authors discussed, only Seeman openly considers the alternative interpretations to mass society theory and attempts to deal with them. In his study conducted with Neal, the concern is with the first two of the three variables — the relationship between the hypothetical independent variable, membership in an association, and the intervening variable, feelings of powerlessness. Although it is not part of their initial design, they devise a clever post facto means of checking on the direction of causality. Unionized workers are compared as to whether they are in open shops, closed shops, or shops which have initial voluntary membership but require an individual to maintain his membership once he has joined. Nonunionized workers are separated according to whether they work for firms where there is no union available or whether they are in open shops in which one can abstain from joining. The purpose of the analysis is to determine whether motivational or structural variables predominate. The results, though quite weak, support both hypotheses. In open shops, where voluntary choice is possible, members are slightly less alienated than are the members of closed shops. On the other hand, all union members (whether members of open or closed shops) are slightly less alienated than nonunion workers.

The authors' findings tend to substantiate the possibility of a symmetrical relationship involving mutual causality. However, caution should be exercised in interpreting even these results. Besides involving very small mean differences, the analysis is unable to control for the fact that there may be very strong informal pressures to join open shop unions — thus limiting the role of voluntary choice based on efficacy — and the fact that variations in either level of education or occupational prestige may actually

be responsible for some or all of the mean differences in powerlessness. This latter possibility appears particularly likely since repeated studies have consistently demonstrated that socioeconomic
status has a strong positive association with feelings of efficacy
(Campbell et al. 1960; Dean 1961). The lower one's socioeconomic
status, the less is his faith in his ability to control his own fate or
to influence political decisions.

Other attempts have been made to specify the nature of the association between alienation and organizational involvement
using solely cross-sectional data, but with equally questionable
results. Erbe (1964) has explored the "self-selection" hypothesis
by asking respondents if they are interested in joining any of a list
of imaginary organizations or if there are any other organizations
they would like to join. The assumption is that if alienation affects organizational involvement, and is thus antecedent to it,
then alienation should be negatively associated with the number
of organizations of interest. Only a slight association is found, leading the author to reject the $A \leftarrow B$ hypothesis. Nevertheless, as
Erbe acknowledges, there are severe limitations to his analysis:
the fact that fictional rather than actual organizational involvement is measured (it being easier for an alienated individual to
say he would join than to actually join), and the fact that *degree* of
active involvement in the organization cannot be evaluated.

In summary, mass society and civic culture theories, which purport to explain the role of secondary associations in affecting political behavior and, ultimately, the political process, have been
"supported" by several major studies based on surveys taken at
one point in time. The interpretation of these data has been chal-
lenged by a series of hypotheses which offer equally cogent alternative explanations for the same empirical relationships. In some
cases the rival hypotheses have been ignored or lost sight of. In
others, attempts have been made to overcome them, though they
have been admittedly crude post hoc tests providing inconclusive
results. The authors most concerned with this problem, Seeman
and Erbe, consequently recommend the use of longitudinal analysis as a means of evaluating the relative strength of the various
propositions. Yet, as is demonstrated by the ITU study, a longitudinal analysis in itself will not necessarily provide an adequate
proof either. What is needed therefore is research which is longitudinal, thus providing the capability of measuring the process of
political socialization *over time,* but which has stronger controls
specifically tailored for considering a problem in which rival ex-

planations are available. One such method is the quasi-experimental approach.

COMMUNITY ACTION: A QUASI-EXPERIMENT

The study to be described here is of a community action agency (CAA), the St. Paul, Minnesota, Head Start program. My purpose is to determine the effect of membership in the program on attitude change. The principal goal in designing the study was to employ a research design that would provide a temporal basis for inferring causality while simultaneously controlling for rival hypotheses. As is common to research conducted in natural, as opposed to laboratory, settings (Kerlinger 1964), true experimentation proved impossible. The self-selection associated with voluntary associations prevents random assignment of cases to experimental treatments. Though individuals presumably are being manipulated by the organizational experience, the process is out of the hands of the researcher. He is limited to systematically observing and recording reactions.

When true experimentation is not feasible, any one of several quasi-experimental designs normally provides second-best alternatives. In such studies, time precedents are exploited and equivalency is approximated—at least on key variables—by matching cases on intuitively important characteristics other than the ones being manipulated. Thus *interrupted time-series analysis* may be used, in which random cases are matched on essential characteristics and then examined as to whether there is a discontinuous change over time in the dependent variables associated with the occurrence of an abrupt, discrete event (chapter 1). Or the panel technique may be used, in which, for the same cases, two or more variables are observed over time to determine whether the occurrence of one or more precedes the occurrence of, or a change in the level of, the other (chapter 2). This may be done using *cross-lagged analysis* (chapter 3), in which the focus is on statistically significant increases or decreases in the relationship between two variables at lagged intervals. When this technique is used, causality is inferred from asymmetry in the cross-lagged correlations. Another alternative is the use of a *pretest-posttest design* involving a nonequivalent control group (Campbell and Stanley 1963:47–50). The latter procedure is used in the St. Paul study and will be described below.

First, however, brief consideration will be given to the choice of the pretest-posttest nonequivalent control group design. In a case such as that of the St. Paul CAA, this design is the best-suited one for the study of organizational effects on individual behavior. Interrupted time-series analysis is inappropriate because it relies on the occurrence of a discrete event which is relatively specific in time (chapter 1). Political socialization, on the other hand, is a continuous process which generally occurs over long periods. Cross-lagged panel analysis, involving large numbers of organizations with repeated replications over time, may be used. However, there are several drawbacks to this technique which make it undesirable in this case. Two will be mentioned here.

First, unless repeated replications are employed (involving considerable expense), cross-lagged analysis cannot cope with the presence of extraneous events or history (chapter 3). The focus solely on cross-lagged correlations between A and B leaves one in a position of being unable to determine whether one variable "caused" a change in the other or whether changes in their association over time are simply spurious (the product of some third factor). In the St. Paul case, the occurrence of the 1970 congressional elections and the rapidly worsening state of the economy were suspected of producing attitudinal and behavioral changes which, without the proper controls, could be confused with an organizational effect.

Second, where synchronous correlations are nonzero and where lagged correlations between A_1 and B_2 are sufficiently larger than those between B_1 and A_2, cross-lagged analysis cannot differentiate between the hypothesis that A_1 causes B_2 and the proposition that B_1 decreases A_2 (Rozelle and Campbell 1969). Similarly, it cannot identify the case where the relationship is in equilibrium, that is, in which A causes increases in B, while B simultaneously produces decreases in A. Since on a priori grounds it was expected that CAA involvement may increase political alienation (Pinard 1968) rather than decrease it, as the mass society thesis predicts, and since I expected that community action programs may attract the more militant and alienated of the poor rather than the more efficacious and satisfied, this limitation also proves a serious drawback.

In short, both interrupted time-series analysis and cross-lagged analysis lack the necessary controls for a study of this type. The pretest-posttest nonequivalent control group design is free of these particular difficulties; and though it too has drawbacks, it is judged the most appropriate for a study of political socialization.

What follows is a basic outline of this design and its rationale, a description of its specific application in the St. Paul study, and a discussion of the types of problems encountered.

A PRETEST-POSTTEST, NONEQUIVALENT CONTROL GROUP DESIGN

Description.   The design used is diagramed as:

$$
\begin{array}{ccccc}
 & & T_1 & & T_2 \\
(O_{A_0}) & X & O_{A_1} & X & O_{A_2} \\
 & & O_{B_1} & X & O_{B_2} \\
 & & O_{C_1} & & O_{C_2}
\end{array}
$$

in which O represents the experimental observations which occurred at two different points in time, T_1 and T_2; X represents the experimental treatment, membership in the organization; and A, B, and C represent the three groups employed in the analysis. Group A includes those individuals who were members of at least several months' standing at T_1 ("old-timers"); group B is made up of the people who were just about to join at T_1 ("newcomers"); and group C is a nonequivalent control group of individuals who were neither members nor about to join at T_1 ("nonmembers").

The design approximates an experiment by taking advantage of a situation where a group of people are being *manipulated* by a government agency. *Temporal analysis* is used to infer causality and control for the rival supposition, attitude $\rightarrow$ organizational involvement. Other rival, methodological hypotheses which threaten internal validity—the effects of testing instrumentation, maturation, and history (Campbell and Stanley 1963)—are partially controlled through the use of the two experimental groups and a *matched* control group. It is assumed that if any one of these factors is operating, the experimental and control groups should react in a similar manner. The design is not precisely experimental because I was merely an observer unable to manipulate the subjects or the treatments to obtain more precise results. More important, since membership is voluntary and a function of self-selection, I was unable to randomly assign subjects to treatment groups and had to rely on matching.

Rationale.   The specific design used in St. Paul consists of several variations of the independent and dependent variables for the purpose of maximizing scope and power of the analysis. The dependent variable consists of seven attitudes involving both personal and political alienation. The attitudes and their measurement will be discussed shortly. There are three variants of the independent variable, organized activity in a voluntary association: (1) *simple membership over a period of relatively short duration* between T_1 and T_2 (six to eight months), involving only the newcomers; (2) *simple membership of longer duration* between T_0 and T_2 (twelve to eighteen months), involving solely old-timers; and (3) *active membership* in the organization's decision-making process, involving a combination of newcomers and old-timers.

The basic logic employed in the design is to observe new members and nonmembers at T_1, to "expose" the first group to X, and then to observe both groups once again at T_2. The purpose is to determine the effect of X on the newcomers, using the unaffected group as the basis for evaluating the significance of any observed change. In this way, if the difference between O_1 and O_2 is significantly larger for the newcomers than it is for the nonmembers, it is inferred that X is the cause. The same procedure is used with the old-timers. However, in this case the organizational effect is assumed to have begun at T_0. Therefore their attitudes at T_1 should reflect any such influence, and the T_1-T_2 interim may be used to examine the implication of long-term exposure to X. Finally, the analysis of active participation in organizational politics is a derivation from the basic rationale of the design. Rather than compare members to nonmembers, active members are compared to inactive members. This makes it possible to determine whether involvement in the organization has any added impact on attitudinal change.

THE SPECIFIC APPLICATION AND PROBLEMS ENCOUNTERED

Three key decisions have to be made in a study of this nature. Each decision entails certain problems. The choice often becomes one of trading one problem for another. A discussion of some of the difficulties associated with relatively short time-series designs of the type used here is presented by Roos in chapter 2 of this book.[3] My purpose is to illustrate these shortcomings in a specific

3. One frequently mentioned limitation of quasi-experimental designs (Caporaso, chapter 1; Roos, chapter 2) is their nontransactional nature. The argument is made that they are unable to handle multiple causality or complex cases in which there is feedback from the dependent to the independent variable. It is true that compared to other techniques (e.g., causal modeling) quasi-experiments are relatively limited in this manner. They are typically used for tests of asymmetrical causality. Nevertheless, it should be stressed that this is a *relative* drawback. Quasi-experiments, I believe, are able to cope with some forms of mutual or simultaneous influence.

setting, as well as the means that are taken to cope with them. Three decisions will be discussed: (1) the choice of organizations to study, (2) the choice of the respondents and matching procedure, and (3) the choice of the specific independent and dependent variables and their measurement. The first decision involves the problem of opting for internal or external validity; the second entails the problem of nonequivalency and the troublesome forms of interaction that accompany it; and the third illustrates the problem of reliability in longitudinal survey analysis.

The Choice of an Organization: Internal vs. External Validity. In designing research a decision often has to be made between internal validity and external validity. In the experimental and quasi-experimental literature the emphasis is placed on the former (Campbell and Stanley 1963:5). Generalizability is sacrificed in order to maximize the ability to control and the ability to detect "true" effects. This was the principal criterion in choosing the St. Paul CAA. The difficulties of revealing a *true* organiza-

---

For example, in the Head Start study I have also focused on the $A \leftarrow B$ link in which attitudes influence organizational involvement. Ideally this may be done at either one or two levels. (1) Based on the interviews conducted at T_1, it can be determined whether attitudes such as efficacy help to differentiate between those who joined the federal program and those otherwise "comparable" respondents who did not. In the Head Start study, however, this actually involved postdiction, since the newcomers to Head Start have already preselected themselves prior to the beginning of the study. More important, one cannot be sure whether differences observed between the newcomers and nonmembers at T_1 are due to some trait which separates joiners from nonjoiners or whether it is a result of the sampling process. Consequently, this technique should only be employed where randomization is possible. (2) A more appropriate opportunity to predict behavior—one used in the St. Paul study—is to compare newcomers among themselves to determine which ones interviewed at T_1 actually become most active in Head Start and which remain little more than members in name only.

Similarly, I am presently exploring the possibility of whether attitude changes induced by organizational involvement can be explained in terms of attitudes originally held at T_1 which may in turn have influenced both members' expectations of the CAA and the likelihood that they will become sufficiently involved to be affected by the experience. Finally, in longitudinal studies such as this, where objective records of participation are kept on a weekly or monthly basis, it should be possible to determine whether individual reactions to the organizational experience affect future levels of involvement within the organization.

The kinds of difficulties I have experienced thus far in exploring the relationships (e.g., the lack of an additional survey at a third point in time and inconsistent record keeping at several of the Target Neighborhood Committees) are of a practical nature. There is no inherent reason why they cannot be considered. I think that though quasi-experiments are restricted in the number of hypotheses they can explore at any one time, it is difficult to specify the exact nature of these limitations. Much of what can be accomplished depends upon the favorableness of the conditions surrounding the study and the availability of supplementary data.

tional effect over time are sufficiently great to warrant the choice of an organization which is likely to manifest an effect, if indeed one does exist. Head Start suits this purpose for three reasons.

First, any one organization is unlikely to have a noticeable added impact on individual orientations among those new members who have been previously active in organized activities. One is therefore more likely to observe a change in behavior among people who, prior to joining the organization, are relatively uninvolved. Though most people are not joiners and do not belong to numerous secondary associations (Hausknecht 1962; Bavchuck and Booth 1969), low-income individuals are even less likely to do so than are members of the middle or upper strata. What is more, they have traditionally been found to possess the traits characteristic of mass society: personal and political alienation, withdrawal from both social and political forms of involvement, a lack of interest or knowledge about the world, limited aspirations, a deep sense of futility and cynicism, intermittent hostility, and a lonely preoccupation with survival (Knupfer 1947; Lewis 1968).

Second, the OEO/HEW doctrine is committed to deliberately creating the types of attitudinal and behavior changes needed to alter the political role of the poor and to encourage them to break out of the cycle of poverty. The goals of CAAs have fluctuated since their introduction in the mid-1960s, and there has been considerable controversy as to their true purpose (Blumenthal 1969). Nevertheless, the dominant ideology has always been one of a "War on Poverty," aimed not at providing more welfare but at destroying the culture which creates and sustains poverty. Today OEO/HEW seeks to uplift the American poor by deliberately working to enhance their sense of self-esteem and self-confidence, to motivate them to help themselves, and to make of them active participants in society.

The technique is one of "community action" and "maximum feasible participation" of the poor in the various government programs which regulate their lives. The assumption is that the deeply rooted economic and political inequalities which produce alienation, despair, and withdrawal cannot be overcome unless the poor are allowed and encouraged to participate in (if not to control) the programs that are meant to serve them. Since the causes of poverty lie in the entire social structure, the poor must be encouraged and helped to participate actively in the political process to seek social and political reform. They must be first politicized, then mobilized and organized as an active interest group to seek those changes within the existing political framework (Donovan 1967; David 1968).

The principal benefactor of OEO/HEW funds has from the beginning been Head Start. The program is designed to give underprivileged children, between the ages of three and five, early training to put them on an equal footing with middle-class youngsters when they begin primary school. Realizing that such children's abilities cannot be developed with just a few hours a day of care and training unless the home environment is also improved, and desiring to use Head Start to implement the broader goals of the fight against poverty, officials seek to involve the parents in the program and in the larger community; to enhance their self-confidence, sense of dignity, and self-worth; and to develop a "responsible attitude toward society" (Office of Child Development 1967, 1970a). In large part these changes are expected to stem from the parents' ability to participate in a democratic decision-making structure. Thus, CAAs are largely bent on doing what Seeman, Almond and Verba, and others believe most organizations can accomplish: eliminating mass society culture. The St. Paul situation presents a good opportunity to determine whether or not they have succeeded in a single instance.

Third, since CAAs are semipolitical organizations, the likelihood of a transfer occurring from the one experience to attitudes about politics in general is greater than in the case of a totally nonpolitical association. It has been argued that latent political functions characterize all groups, whether they are of a directly political nature or not, and that any organization should be able to reduce alienation (Greer and Orleans 1962). The data are unavailable to either adequately support or deny the contention, but it would seem that with studies of this sort a beginning should be made with the more plausible case and then proceed from there.

The costs of maximizing internal validity at the expense of generalizability are very real. The extent to which conclusions drawn from the St. Paul study can be applied to an understanding of organizational effects in general is limited. Very little can be inferred about mass society in general, though careful conclusions may be drawn about the influence of organizational membership on behavior. The 1970–71 St. Paul antipoverty program is sufficiently like the majority of other community action programs, where the emphasis is on social service and *individual change,* to allow for the valid transfer of these findings to most CAAs (Rose 1972). Nevertheless, the findings are probably only poorly applicable to those CAAs which primarily seek to induce political mobilization and *institutional change,* a type which flourished briefly during the initial months and years of the War on Poverty (Moynihan 1969). Considerable caution must be employed when attempting to compare these findings to other types of organizations and to those having a different kind of clientele. They may prove most useful, however, when applied to the political development and mobilization literature dealing with the organization and amelioration of the masses in the underdeveloped areas of the world (DeTocqueville 1955; Deutsch 1961; Huntington 1968). As is true with much of the research in political science, a full appraisal of the external validity of the study must await its comparison with the results of similar efforts now available and those yet to be made.

The Choice of Respondents, Equivalency, and the Problems of Selection-History and Selection-Regression Interaction. The St. Paul respondents compose a panel which was initially interviewed in the fall of 1970 and then reinterviewed in the spring of 1971.[4] The control group was matched as closely as possible to the experi-

---

4. There were 250 children enrolled in the St. Paul program; each child daily attended one of the eleven centers. Of these, an effort was made to interview the female parent of each of the children in five selected centers. The decision to interview only the female parent was based on the large proportion of Head Start children who come from fatherless homes, the fact that in previous studies (Campbell et al. 1960) female respondents have typically indicated lower feelings of political efficacy than males, and the desire not to have to add a further control for sex in a study which was limited from the outset by a relatively small N.

Centers were chosen on the basis of several known characteristics, these being their reported levels of *prior parent involvement, ethnic composition,* and the *proximity of the respondents to the centers.* Two of the five centers were considered by Head Start officials to be generally successful in terms of getting parents interested and active in the program. One was considered to be moderately successful and two relatively unsuccessful. As it turned out, during the period of the study parents at the five centers volunteered on the average of five and one-half hours of service per month, ranging from one center in which no one donated the required "minimum" of eight hours per month to another in which 44 per cent contributed the minimum or more. The average who actually attended one or more Target Neighborhood Committee meetings is 50 per cent, ranging from a low of 24 per cent at one center to a high of 62 per cent at another. The average number attending one or more Policy Advisory Council meetings is 17 per cent. Those who did attend were predominantly elected delegates who were obligated to participate. In all but one center the membership is over 50 per cent white, the exception being 70 per cent Mexican-American. The number of Negroes enrolled ranges from 0 to 30 per cent. Only those centers which service a relatively small geographical area were chosen for study, so as to eliminate as much as possible the practical problems of transportation costs and traveling time, which serve as barriers to parental participation. Three of the five centers serviced densely populated public housing projects; the remaining two are located in low-income residential neighborhoods.

mental group.[5] Its members were selected randomly from a list of St. Paul ADC mothers who live in the immediate locality of the Head Start parents and who have one or more children of the same age group (three to five). The mortality rate for the panel over the six- to eight-month period (depending upon when an individual joined) is 20 per cent. With the omission from the analysis of 10 experimental subjects who quit Head Start during the interim, there remain 82 members, consisting of 48 newcomers and 34 old-timers, and 31 nonmembers who were successfully interviewed both times.[6]

The goal of quasi-experimental designs is to approximate equivalency through matching when pure randomization is impossible. The Head Start study illustrates the utility and the limitations of this technique. If done properly, all matching guarantees is that the experimental and control groups will be similar on the *specific* characteristics upon which the matching is based. There may remain crucial differences which were not originally foreseen by the researcher. Thus in the St. Paul study, my principal purpose was to create a control group which shared the same *life style* as the Head Start membership. On this point I was generally successful (table 7.1).

Members of both groups have incomes at or below the poverty level, have young children of the same age, live in the same neighborhood, and are roughly similar in ethnic background, race, and class identification. Less than 25 per cent of both groups reported themselves as being the chief wage earner for their families; all received some form of welfare assistance, and over half dropped out of school before finishing the twelfth grade.

Yet several outstanding differences remain. Members of the control group have lived at their present address for a longer period of time, are older, and are much less likely to be married. The difference in marital status is clearly a direct result of sampling ADC mothers.[7] Whether the differences in length of resi-

5. The closeness is suggested by the fact that six original control respondents joined Head Start during the interim and had to be switched to the experimental group.

6. Only a small minority of those whose attitudes were not recorded at T_2 refused to be interviewed. There are too few of them or of the quitters to systematically draw any conclusions explaining their actions. The reasons given to the interviewer varied from case to case.

7. With the aid of hindsight, it appears that a more appropriate procedure would have been to draw the control sample from a population of welfare recipients in general. The decision to use ADC lists was based on the assumption that the Head Start clientele consists primarily of ADC children.

TABLE 7.1
COMPARATIVE LIFE STYLES OF HEAD START NEWCOMERS AND NONMEMBERS

Life Style	Newcomers (percentages)	Nonmembers (percentages)
Subsisting on a *poverty income* or less	100	100
One or more *children between 3 and 5 years* of age	100	100
Residing in the *vicinity of a center*	100	100
Residing at *present address* for one year or more	51	77
Residing in *St. Paul* for one year or more	90	95
Parents *native born*	82	77
Race		
White	61	73
Black	17	10
Mexican	22	18
Class Self-Identification		
Lower	18	19
Working	38	32
Middle	26	32
Other/DK	18	16
Age of Respondent		
20–26	57	28
25–30	20	28
31 or over	23	44
Currently married	49	16
Education		
Public school dropout	51	51
High school graduate	28	35
Post-high school training	21	14
Chief wage earner	22	21

dence and age was a further result of marital status is undetermined.

Though unfortunate, the lack of greater equivalency between groups is not necessarily as crippling as it is simply disappointing. Even without complete equivalency, members of both groups can be said to share a very similar life style. What is more important, controls are possible for the rival $A \leftarrow B$ hypothesis as well as for the effects of testing, instrumentation, and maturation — in short, for most of the common alternative interpretations. Regardless of the differences that do exist, the control group, not having been exposed to X, should remain relatively unchanged during the interim.

There remains, however, the more complicated matter of selection-interaction. In the St. Paul study this presents two principal problems: that of *selection-history* and that of *selection-regression interaction*.[8] The possibility of selection-history interaction is due primarily to the differences in marital status. Since married women are likely to be influenced by their spouses, it is possible that differences in the reaction of the sexes to extraneous variables during the interim could produce results which are then confused with organizational effects. One can control for this eventually, however, by holding the suspected variable constant during the analysis stage.

The second form of interaction, that between selection and the statistical artifact of regression toward the mean (Campbell and Stanley 1963:10–12), presents a problem solely in the analysis of organizational participation involving the comparison of non-activists and activists. The fallacy commonly associated with this type of interaction is to confuse an experimental effect with random changes from extreme scores toward the group mean. Though steps are taken to circumvent this problem in the final section of the analysis, they are only of limited success.

Attitudes, Involvement, and Their Measurement: The Problem of Reliability. During the interim between the two surveys, parents in the St. Paul Head Start program were strongly urged to participate in monthly meetings of the citywide Policy Advisory Council (PAC) and local Target Neighborhood Committees (TNC), as well as to volunteer as aides in the classrooms. The members who participated in organizational politics by attending the meetings

---

8. A related problem of the St. Paul study, one not peculiar to designs of this nature in general, is relevant in this context. There is interaction between the time of the interview at T_1 and an extraneous variable, campaign effects. Interviewing during the first survey was strung out over a period of two and one-half months as new members trickled into the program. The 1970 congressional and statewide elections occurred during the end of this period, and there is some evidence that it did affect responses. It is because of the real possibility of a campaign effect biasing the initial interview that steps are taken in the subsequent analysis to circumvent a possible contamination of results. First, changes occurring between the first and second survey are examined as to their relationship with *degree of involvement* in the CAA. If it can be demonstrated that the greater the level of participation, the greater the likelihood of change, then it may reasonably be inferred that such change is a result of organizational involvement—unless, that is, there is some reason to believe that there is interaction between the campaign effect and the level of PAC/TNC attendance. There is no basis for expecting this type of interaction. Second, in those cases where the effect of membership alone is considered (regardless of level of CAA involvement), *select groups* are used, composed solely of people who were interviewed during a two-week period just prior to the traditional Labor Day kickoff for political campaigns.

were predominantly the same people who volunteered at the local centers. Objective weekly and monthly records were kept of attendance at both types of activities. In the following analysis, organizational "involvement" and "participation in organizational decisions" will be inferred from frequency of attendance at the PAC/TNC meetings.

Since the purpose of the analysis is to determine whether CAAs are able to reduce both personal and political alienation, the respondents' general sense of both personal control (Gurin et al. 1969) and political efficacy (Campbell, Gurin, and Miller 1954) are measured. In addition to personal control or self-confidence, two additional forms of *personal* alienation — anomie (McClosky and Schaar 1965) and self-esteem (Rosenberg 1965) — are included in the analysis. Both have been discussed in terms of mass society (Kornhauser 1959; Scrole 1956). It is assumed that as a result of the absence of organized activities and the opportunity for social participation in mass society, an individual experiences little sense of social purpose. This adversely affects one's opinion of his own worth and ultimately encourages self-estrangement. Anomie, defined here as the extent to which an individual is confused about events going on around him and what is expected of him by others, is accredited in part to social isolation, including a lack of organized involvement. Associational membership is alleged to remedy such ailments by providing people with a sense of purpose, self-confidence, a steady flow of information as to what is appropriate behavior, and the security that accompanies well-developed social ties.

In order to estimate the potential breadth of organizational effect, three indicators of *political* powerlessness are used: (1) the political efficacy scale (Campbell, Gurin, and Miller 1954), which measures a person's perceptions of his ability and the ability of people like himself to influence public officials; (2) the sense of popular control index (modified from the control ideology index of Gurin et al. [1969]), which indicates perceptions of public officials' responsiveness to people in general; and (3) a modified version of the system modifiability index (Gurin et al. 1969), which assesses the person's belief in the possibility of system reform. System modifiability items were only asked of people who indicated cynicism in responding to popular control questions. Used in conjunction with the other indices, they make it possible to talk about the degree of alienation by determining whether those who do feel powerless are so forlorn as to give up any hope that things can be improved. Finally, in order to tap the related area of general affect toward government, a government evaluation index is employed, based on respondents' answers to questions asking them to evaluate governmental performance in the handling of specific problems designated by respondents themselves (in open-ended questioning) as being important.

A serious question which must be faced by those using quasi-experimental design involving survey research is whether or not social scientists have perfected sufficiently sophisticated and precise attitudinal scales and other interview items for conducting longitudinal analysis. In the case of the St. Paul research, this is a particular problem, considering (1) the excessive length (approximately eight months) of the study, (2) the nature of the sample (low-income, poorly educated respondents who are less likely to be characterized by cognitive restraint and well-developed informed opinions, and who are less able to comprehend questions asked of them [Converse 1964]), and (3) the fact that an attempt is being made to alter both attitudes and behavior, thus confusing the question of reliability with deliberately induced change. To circumvent this problem, scales of demonstrated reliability were used wherever possible (Robinson, Rusk, and Head 1968). In addition, a small pretest was conducted to determine the prospective respondents' level of comprehension, the result of which is a simplification of the language employed in several of the original scales. Despite these efforts, the test-retest reliability coefficients are not impressive,[9] suggesting either that opinions changed over time as a result of some factor such as the experimental treatment or history, or that the interviewers often measured what Converse calls "nonattitudes" (1963), a goodly proportion of the answers being purely random responses to the stimulus of the interview.

Poor reliability increases the likelihood of a regression effect (Campbell and Stanley 1963). It also impairs the power to reject

9. Test-retest correlation coefficients for nonmembers and members respectively are: personal control, .46/.48; self-esteem, .32/.41; anomie, .50/.43; political efficacy, .52/.33; belief in popular control, .51/.38; government evaluation, .24/.25; and system modifiability, .09/.37. Since both groups experienced similar attitudinal changes during the interim — apparently as a result of the declining health of the economy and postcampaign effects — part of the low reliability may be explained in terms of a true opinion change. In the case of the members, additional change may also be accredited to organizational effect. That a proportion of the variation reflects the recording of nonattitudes among poorly informed individuals is indicated by a generally higher level of reliability for well-informed respondents. For example, the two incidents of lowest reliability are nonmembers' beliefs about governmental performance (.24) and system modifiability (.09). Among just the better-informed nonmembers, however, the test-retest coefficients increase considerably, to .59 and .58 respectively.

the "no organizational effect" hypothesis. Though faulty items may result in artificial variations in responses over time, such results, being random, are readily observable with the use of both the control group and tests of statistical significance. However, to the extent that real variations in orientation actually occur as a function of involvement in the organization, they may go undetected because of the inability of crude instruments used with relatively unsophisticated respondents over a prolonged period of time to measure anything less ⁺han gross behavioral or attitudinal changes. To the extent that no organizational effect is found, it may be due to the presently limited ability to record real changes which may have occurred. To the extent that such change is observed, there will remain the question of how much greater it might have been had we had available more sensitive and precise instruments.

ANALYSIS OF RESULTS

The analysis is divided into four sections, each evaluating the mass society thesis from a different vantage point. The first section focuses solely on the effect of membership on those individuals entering the program for the first time. The second section turns to the long-term effects of membership by incorporating the old-timers into the analysis. Attributes of CAAs which contribute to change are discussed in the third section. The final section is specifically concerned with the civic culture variant of the mass society thesis, in which active participation in making organizational decisions is of prime importance. None of these analyses substantiates the mass society thesis. The first three contradict it while the fourth provides only very questionable support.

THE EFFECTS OF MEMBERSHIP

The analysis of the effect of simple membership in the St. Paul Head Start program reveals only slight influence on psychological orientations toward oneself or the political system. Much of the difference existing between the nonmembers and the newcomers at T_2 is the result of self-selection, not political socialization. The organizational effect that is observed is in the direction of greater, rather than less, alienation.

Self-Selection vs. Organizational Effect. Newcomers to Head Start reveal attitudinal changes (indicated by $\overline{D}$ in table 7.2 for the mean

TABLE 7.2

A Comparison of Mean Changes ($\bar{D}$) Between T_1 and T_2 of Newcomers' and Nonmembers' Attitudes toward Self and System *

	Newcomers				Nonmembers				Significance of the difference between $\bar{D}$s
	T_1	T_2	$\bar{D}$	Sig. of $\bar{D}$†	T_1	T_2	$\bar{D}$	Sig. of $\bar{D}$‡	
Self-evaluation									
Personal control	12.22	12.72	.50	—	10.74	11.41	.67	(t = 1.05)	—
Self-esteem	19.83	21.33	1.50	(t = 1.60)	18.57	19.86	1.29	(t = 2.24)	—
Anomie	4.78	5.28	.50	(t = 1.04)	6.87	6.52	−.35	—	(t = 1.09)
System evaluation									
Political efficacy	9.33	9.39	.06	—	9.83	9.44	−.39	—	—
Popular control	3.61	3.11	−.50	(t = 1.09)	2.65	2.48	−.17	—	—
Government evaluation	4.93	3.40	−1.53	(t = 2.25)	4.77	3.83	−.94	(t = 1.28)	—
System modifiability	3.25	2.92	−.33	—	3.31	3.40	−.09	—	—

* Only a selected group of newcomers and nonmembers are analyzed here: the newcomers ($N = 18$) and nonmembers ($N = 23$) who were initially interviewed during the two weeks just prior to Labor Day.

† Significance at .05, using a one-tailed t test, requires a $t \geq 1.74$ for newcomers and 1.71 for nonmembers (small ts not indicated).

‡ Significance at .05, using a one-tailed t test, requires a $t \geq 1.68$ (small ts are not indicated).

differences between T_1 and T_2 scores) over the eight-month period, but the changes are closely matched by those of the nonmembers. In several cases the changes experienced over time by one or both groups are statistically significant (self-esteem and government evaluation), but in no instance is the difference *between* the mean changes significant. Thus, though there is some statistical basis for concluding that something other than random variation has occurred here, there is not sufficient justification for assuming that it has anything to do with being a member of Head Start.

It is interesting to note that if I had based the analysis solely on data collected at T_2, in which I compare current members in an organization with nonmembers, I would have drawn conclusions similar to those of the mass society theorists – a conclusion which, at least in this case, would be invalid. This is clearest with regard to self-evaluations. At T_2, new members are less personally alienated than nonmembers. They indicate a higher sense of personal competence (12.72 and 11.41, respectively) and of self-esteem (21.33, 19.86), while having a lower sense of anomie (5.28, 6.52). Superficially it would appear that Head Start had the predicted organizational effect. An examination of T_1 responses, however, reveals that the newcomers were less personally alienated *before* they entered the program. Similarly, though more ambiguous findings exist for system evaluation, with one exception (government evaluation) those differences that are observed at T_2 existed *prior* to the new members entering Head Start. The analysis therefore supports the rival $A \xleftarrow{(+)} B$ hypothesis that the reason for a positive association between personal evaluation and activity is that the less personally alienated individuals are more apt to join voluntary associations.

There is, however, the alternative hypothesis that the difference between the two groups is due simply to the failure to obtain complete equivalency in the sampling process. Further analysis was therefore conducted to determine the merits of this interpretation. A check was made of the differences in levels of alienation associated with marital status – the principal difference between the two groups being that nonmembers (who were selected from ADC lists) are much less likely to be married. The results reveal that, within both the experimental and control groups, married women are more likely to feel both politically and personally powerless.[10]

10. The greater alienation associated with married life probably reflects male influence on the respondents' political orientations. Husbands have traditionally played a more influential role in American politics and in family discussions about

Since the Head Start women as a group are more likely to be married, yet less personally alienated, differences produced by the sampling procedure serve to *conceal* rather than to produce the differences observed between nonmembers and newcomers to the organization. Much of the T_2 difference between the two groups therefore is a product of self-selection, not socialization.

Community Action and Alienation. Though none of the mean changes experienced by the experimental group prove significantly different statistically from the changes among the control group, there is one interesting tendency which may have important meaning in that it suggests the need for further elaboration and development of mass society–civic culture theories. There does appear to be a slight to moderate organizational effect, but it is an *adverse* one. Instead of alleviating political alienation and anomie, membership in the CAA appears to have aggravated both. Specifically, during the interim new members became more bewildered by events around them and confused as to what was expected of them, while nonmembers decreased in anomie. In addition, the newcomers reveal a greater, but nonsignificant, tendency both toward a weakened faith in the effectiveness of popular control over elected officials and the possibility of system reform and toward a lower opinion of governmental performance.

The crucial question, of course, is whether these differences really indicate an organizational effect, whether they are a result of measurement effects or some other exogenous variable which influenced both groups independently,[11] or whether they reflect simple random fluctuation. The latter is quite probable, since in only one of the cases, that of anomie, does the difference between the mean changes of the two groups approximate statistical significance. The question of measurement bias may be ruled out, since both groups experienced the same influences; they were asked the same questions, by the same interviewers, during the same time period. There is reason to believe, however, that de-

politics (March 1953), including decisions as to whether to vote (Glaser 1959) and the nature of partisan support (McClosky and Dahlgren 1959). Since in American culture it is also primarily the husband's duty to provide for and support the family and since most of the male spouses in the study are unemployed, it may be that the males feel a greater humiliation and lack of self-confidence for their failure to live up to their culturally defined roles, as well as a greater sense of political cynicism and powerlessness, and that these feelings are passed on to the wives.

11. The relationships persist when marital status is controlled. A comparison of nonmarried members and nonmembers over time indicates a greater adverse change among the newcomers.

spite the slightly greater tendency toward political alienation on the part of the newcomers, the same tendency among *all* respondents primarily reflects the effect of *history*.

The Head Start study was conducted during a period characterized by simultaneous inflation and economic recession. Prices were rising month after month, while more and more people were being laid off work. The people an ailing economy is most likely to affect are the poor, who have fewer dollars to spend or savings to fall back on, those living on fixed incomes (whether they be old-age benefits or welfare checks), and those most likely to be unemployed—the unskilled and poorly educated. These are characteristics of all the respondents in the Head Start study, and it is understandable that members and nonmembers alike experienced greater dissatisfaction with, and cynicism toward, the political system for its inability or refusal (depending on how they saw it) to provide jobs and control inflation. The extent of their dissatisfaction is most clearly visible in their appraisal of governmental performance. Respondents had significantly lowered their evaluation of it by the time of the spring survey. The principal type of problem volunteered by the respondents when asked their opinion of governmental performance was economic in nature (including primarily an inability to find jobs and the high cost of food and rent).

In summary, the St. Paul CAA has little impact on the attitudes of new members during a period of approximately half a year. The differences that do exist between new members and nonmembers are largely due to self-selection, not socialization. The only discernible effect of the program is an adverse one which heightens anomie and aggravates political alienation. Even here, however, the presence of an organizational effect is questionable. Only the differences in the mean changes in feelings of anomie approach statistical significance. The increase in political alienation primarily takes the form of an intensified dissatisfaction with government services. The fact that it, as well as the milder increases in feelings of political futility, occurs for both groups suggests that the cause is principally *history*, not the CAA.

MEMBERSHIP OF LONGER DURATION

It is nevertheless plausible that the greater increase in sense of anomie and the slightly greater increases in political powerlessness found for newcomers are the signs of a weak organizational effect—perhaps one that becomes more potent over periods of longer duration. If this is the case, it should be reflected in the attitudes of the roughly 40 per cent of the membership who are old-timers – those individuals who had been in the organization up to a year (membership is limited to two years) before the newcomers ever joined. Old-timers should be more anomic and politically alienated at T_1 than those individuals who have not yet been exposed to the organization, and they should continue to suffer adverse reactions between T_1 and T_2 during the second year of membership.

Newcomers and Old-timers at T_1. A comparison of incoming and existing members at T_1 reveals the same heightened anomie and weak tendency toward political alienation found in the T_1–T_2 analysis of newcomers. In addition, the effect of longer duration appears to have produced a deterioration in the self-image of the older members (table 7.3).

The old-timers are consistently, but only slightly, more *politically alienated* on each of the indices than the individuals about to enter the organization. Once again, there is a nearly significant difference for anomie. Thus, in this respect, the data conform very closely to the T_1–T_2 analysis of newcomers and nonmembers. In neither case is there any type of support for the mass society thesis. If anything, the fact that two successive generations reveal the same apparent negative reaction to the program suggests that the

TABLE 7.3

A COMPARISON OF MEAN SCORES OF NEW AND OLD MEMBERS AT T_1*

	Newcomers $(\overline{X})$	Oldtimers $(\overline{X})$	Significance of the Difference †
Self-evaluation			
Personal control	12.22	11.50	—
Self-esteem	19.83	18.72	$(t = 1.09)$
Anomie	4.78	6.07	$(t = 1.48)$
System evaluation			
Political efficacy	9.33	8.81	—
Popular control	3.61	3.04	—
Government evaluation	4.93	4.73	—
System modifiability	3.25	3.09	—

* Only a select group of newcomers and old-timers are analyzed here: the newcomers ($N = 18$) and old-timers ($N = 27$) who were initially interviewed during the two weeks just prior to Labor Day.

† Significance at .05, using a one-tailed t test, requires a $t \geq 1.68$ (small ts are not indicated).

experience of community action is to induce rather than alleviate alienation.

The picture is compounded and faith in the mass society thesis still further diminished by an unexpected discrepancy between old-timers and newcomers. The former are more *personally alienated* at T_1. It has been seen that membership is associated with a heightened sense of anomie for newcomers; other than that, however, they show no sign of being affected in terms of their own self-evaluations. Yet, in the fall of 1970, old-timers not only were more likely to experience a high level of anomie, but they had a lower sense of self-confidence and self-esteem as well. Though only the differences for anomie and self-esteem approach statistical significance, all three are the *opposite* of that predicted. If it is assumed that a change has occurred between T_0 and T_1 for the old-timers and that the differences are not due solely to chance (tenuous assumptions at best), the only conclusion possible is that the long-term impact of the program on self-evaluation is completely the opposite of that intended by OEO administrators or preconceived by mass society theorists. Those who enter CAAs may initially be more self-confident and self-assured, but if the St. Paul Head Start program is any indication, the effect of their experiences in the association may be to cause such feelings to deteriorate.

The Old-timers during T_1–T_2. The likelihood of enhanced personal estrangement, as well as some form of political alienation, is corroborated by a comparison of the mean changes ($\overline{D}$s) during the T_1–T_2 interval (table 7.4). Like the newcomers, the old-timers become more anomic. Though in the fall they already revealed a greater level of confusion and insecurity than the new people entering the program, they continued to become even more perplexed over the extended period of time. In contrast, it will be recalled, nonmembers decreased in anomie. In addition, despite the propensity of both new members and nonmembers to experience roughly similar feelings of self-confidence and self-worth, the members of longer standing showed signs of greater self-diminishment. Thus long-term membership offers no suggestion of a reduction in self-alienation. In each case—anomie, self-esteem, and self-confidence—nonmembers became less alienated while old-timers either stayed the same or became more alienated.[12]

12. The difference in mean changes ($\overline{D}$s) for old-timers and nonmembers approaches statistical significance in each case: anomie, $t = 1.35$; self-esteem, $t = 1.26$; personal control, $t = 1.07$.

TABLE 7.4

A Comparison of Mean Changes ($\bar{D}$) between T_1 and T_2 of New and Old Members' Attitudes toward Self and System *

| | Newcomers | | Old-timers | | Significance of the Difference |
	$\bar{D}$	Significance of $\bar{D}$†	$\bar{D}$	Significance of $\bar{D}$†	between $\bar{D}$s ‡
Self-evaluation					
Personal control	.50	—	−.39	—	—
Self-esteem	1.50	($t = 1.60$)	−.09	—	($t = 1.24$)
Anomie	.50	($t = 1.04$)	.78	($t = 1.32$)	—
System evaluation					
Political efficacy	.06	—	.41	—	—
Popular control	−.50	($t = 1.09$)	−.63	($t = 1.43$)	—
Government evaluation	−1.53	($t = 2.25$)	−1.46	($t = 3.10$)	—
System modifiability	−.33	—	.26	($t = 1.23$)	($t = 1.54$)

* Newcomers = 18; old-timers = 27.

† Significance at .05, using a one-tailed t test, requires a $t \geq 1.74$ for newcomers and 1.71 for old-timers (small ts are not indicated).

‡ Significance at .05, using a one-tailed t test, requires a $t \geq 1.68$ (small ts are not indicated).

If the long-term picture is dim with regard to the impact of CAAs on self-evaluation, it is only slightly brighter for system evaluation. Old-timers are as likely as newcomers to become increasingly cynical about the general responsiveness of government officials to popular will, and they become equally dissatisfied with government efforts to cope with community and national problems. The basis for any optimism rests with the weak propensity of old-timers to increase in both political efficacy and faith in the feasibility of system reform, while the nonmembers decline. Together this suggests that perhaps, eventually, heightened mistrust and dissatisfaction are moderated by the hope that conditions can be amended through political action. Yet both of these relationships are weak and quite likely due to chance.[13]

In summary, the conclusions about both short-term and long-term membership in the St. Paul Head Start program are the same. There is little basis for accepting the mass society thesis that a grass-roots voluntary association will create favorable attitudes toward the system or self. *What little effect is evident is predominantly negative, contributing to anomie, political alienation, and, in the case of long-term membership, self-estrangement.* With the exception of anomie, even these changes are generally weak and of questionable significance.

DISCONTINUITY IN THE SOCIALIZATION PROCESS

The weak tendency toward an adverse effect, however, suggests the need for a rethinking of the mass society thesis. As it is presently formulated, the thesis rests primarily on one aspect of organizational membership: the hypothetical propensity of people to identify with the organization and thereby to obtain confidence from its collective power, as well as a sense of purpose from its goals. Organizations, though, are both complex and varied. There is no reason to suspect that they all will have the same effect — even those which are both voluntary and democratic in nature. Assuming that the mass society theorists are correct in their assumption about the phenomena of identification, within any one association there may be multiple forces at work, some of which cancel out or overwhelm those influences contributing to favorable personal and political attitudes.

Clearly, in the light of the St. Paul findings, considerable

---

13. In no case are the differences between old-timers and nonmembers regarding political attitudes significant, nor do they approach significance.

thought could be usefully applied to classifying organizations according to their types and the effects they produce, and to exploring the various facets of organized behavior which contribute to the socialization process. It would appear, for example, that the nature of the socialization process will vary depending not only on the nature of the organization per se (its structure, goals, etc.) but on the nature of the membership. There is sufficient evidence to suggest that organizations composed primarily of deprived citizens, particularly CAAs, are likely to produce, rather than to alleviate, alienation. I can best illustrate this point by briefly elaborating on what I believe to be the effects of two organizational processes — *enculturation* and *communications* — on the poverty-stricken members of CAAs.

Enculturation. There is a growing store of information on the politics of the poor and community action which indicates that CAAs are commonly associated with increases in dissatisfaction and active protest among the poor. An antiestablishment atmosphere has generally characterized many of the programs ever since their introduction in the mid-1960s (Donovan 1967; David 1968; Blumenthal 1969). Particularly during the first few years of their existence, they were aimed principally at arousing the poor against established political, welfare, and educational institutions which tend to subjugate, humiliate, and discriminate against lower-income groups and racial minorities. As a result, CAAs have been accused of (or praised for, depending on one's point of view) having stimulated the poor to actively dissent and to participate in the political process in order to make known their needs and demands for remedial action.

Scattered research indicates that there is truth to these claims, but that the extent of the organizational effect varies considerably depending on the nature of the CAA (Venecko 1969). The St. Paul Head Start program is by no means a militant CAA of the type located in other communities; but even in those CAAs where outright protest is played down or ignored altogether, individuals are still apt to be taught that the guilt for their poverty rests in part with the nature of the existing system, and that they — the poor and the nonwhite — are the victims of society and are powerless in the face of forces over which in the past they have had no control. The purpose of the indoctrination is to restore self-confidence and self-respect to these people by transferring the blame for failure from the poor to the system, as well as to sell community action itself as a means of obtaining greater political influence. In the

St. Paul case, such efforts failed and, if anything, aggravated personal alienation over the long run. However, to the extent that such goals are sought by placing blame for failure with the system (e.g., accusations of racial discrimination, inequitable representation, etc.) rather than with personal ineptitude, they are likely to encourage a certain amount of political cynicism and hostility as well.

Communications. A second, more general, explanation of the tendency for Head Start membership to result in heightened political alienation lies in the nature of the membership and the centralized communication network available in most organizational settings. Organizations enhance communications, facilitate the flow of information, and thus increase both the saliency and comprehension of issues and one's ability to deal with them. A variety of studies have demonstrated that the more socially or politically involved an individual is, the greater is the likelihood that he will be well informed, opinionated, and sophisticated in his understanding of the political system (Key 1964; Converse 1964; Mc-Closky 1964). The Head Start study is similar in this respect. The mean level of political information of old-timers and newcomers at T_1 is 5.44 and 3.67, respectively; the mean increases in level of information of both newcomers and nonmembers during the interim is 4.72 ($t = 3.66$) and 3.30 ($t = 2.46$). Thus membership is associated with a heightened political awareness.

I will refer to this effect as *reality therapy*—the process by which an individual learns about himself, the world, and his place in it. It is useful to make an analytical distinction between it and the enculturation process, referred to above, in which myths, cultural images, and other, possibly distorted, information is transmitted to the individual from organizational officials or from one's peers. The effect of community action therefore may be looked at in terms of the extent to which it acts upon previously isolated, uninformed, and politically naïve individuals by providing them with information as to the nature of society and their place in it.

What, then, are poorly educated, poverty-stricken, and predominantly unemployed individuals likely to learn in an organizational context? They will discover (if they do not already know) their lowly status in society. They are likely to learn that in fact the responsibility for their failure to succeed in life does not rest entirely with them, that some people begin life with immense advantages while others are trapped in a culture of poverty or by racial and ethnic barriers which are beyond their control (Knupfer

1947). They are likely to become increasingly familiar with the realities of politics, while becoming acquainted with ways to gain access to public officials and acquiring tactics of political influence.[14] However, they will also discover stubborn bureaucracies, red tape, competing pressure groups, budget problems, and empty promises. They will, in short, learn that politicians and the political process can be influenced, but that it is a difficult and often a very frustrating process and that they, the poor, are generally not in a favorable position to possess such influence (Cloward and Piven 1971). Such appears to be the case in St. Paul, where the well-informed respondents are the most politically alienated. *For the very poor, therefore, knowledge is often a source of alienation.*

A specific illustration of the ability of an organized setting to facilitate the flow of information and, in so doing, to adversely affect the system evaluations of the poor is made possible by an accidental occurrence during the initial survey. The incident, though unique, offers some insight into the process by which organizational and other forms of involvement facilitate poor people's awareness of their relative powerlessness. In the third week of August, at the end of the first week of interviewing, the mayor of St. Paul unexpectedly slashed the city's welfare budget by $2.3 million. Several days later, when the members of the Welfare Board objected and voted to reinstate the funds, they were fired en masse. Since all of the subjects of the study are welfare recipients, an adverse reaction could well be expected. However, to the extent that Head Start serves as a communications hub, disseminating information and providing a place to discuss it, the impact among the membership should be even greater than among the nonmembers.

To test this proposition, those existing Head Start members (the old-timers) who were interviewed the week before these events occurred were compared to the members interviewed the week following. The same was done with nonmembers. All four groups were then compared on a set of attitudes which could be expected to be affected by the incident (table 7.5). In addition to the other variables which I have been using throughout the analysis, two additional forms of system evaluation are included: sense of administrative efficacy, which is derived from respondents' statements concerning their ability to handle hypothetical traffic and administrative problems, and a sense of police efficacy, similarly

14. In the St. Paul Head Start program, for example, parents were introduced to mass letter-writing campaigns, attendance at city council meetings, visits to state officials, legal aid and related services.

TABLE 7.5

A Comparison of Mean Scores of Those Members and Nonmembers who were Interviewed the Week Prior to the Welfare Incident and the Week After

	Existing Membership (N = 34)			Nonmembership (N = 34)		
	Interviewed Before (N = 21)	Interviewed After (N = 13)	Difference	Interviewed Before (N = 12)	Interviewed After (N = 22)	Difference
Personal control	12.05	10.31	-1.74 (p < .10)	11.50	11.68	.18
Political efficacy	9.70	6.69	-3.01 (p < .01)	9.75	8.14	-1.61 (p < .05)
Administrative efficacy	8.52	6.54	-1.98 (p < .10)	8.33	8.32	-.01
Police efficacy	6.81	4.85	-1.96 (p < .05)	6.92	6.32	-.60
Popular control – general	2.48	1.31	-1.17 (p < .05)	1.33	1.86	.53
Popular control – subgroup	.67	.23	-.44 (p < .10)	.33	.96	.60 (p < .05)
Government evaluation	4.44	4.60	.16	5.25	4.29	-.96
System modifiability	2.94	2.92	-.02	3.40	3.38	-.02

based on an individual's perceived ability to cope with a case of mistaken identity by city police. Sense of popular control was recorded at that time as two different variables—the political control of people in general and the political control of unidentified subgroups.

With the exception of government evaluation and system modifiability—where no significant changes were recorded—the reaction to the welfare cut and mass firing of the board is much more severe among the Head Start members.[15] In each case (personal control, political, administrative, and police efficacy, and popular control) a significant or near-significant drop in efficacy occurred between those members interviewed just before and immediately after the event. Among the nonmembers, a similar reaction occurred only with respect to their sense of political efficacy; even here, the magnitude of the drop is dwarfed by the comparable change among the Head Start membership (-1.61 to -3.01). A reasonable inference from this data is that as a result of their involvement in Head Start—an organization in which such events are both salient and central—members were much more affected by the St. Paul budgetary politics of 1970–71 than were others who lived more isolated and sheltered existences. As is so often the case, the less organized, more disadvantaged groups in society fared poorly in that process, with the result that those who became aware of the incident developed an even greater sense of futility and frustration. In this particular case, the communications process appears to have reinforced the development of political alienation. A similar form of interaction between organizational effect and history was probably operating in St. Paul during the T_1–T_2 interim as well. The apparent tendency of members to react more adversely to the sagging economy than did nonmembers may reflect the greater saliency of the subject in the organizational setting.

Additional Evidence. Evidence provided by several other studies indicates that this type of reaction is not unique to the St. Paul CAA. A nineteen-month study conducted in Topeka, Kansas, involving personal observations of parent participation in twelve neighborhood action committees reports that, in the early stages of the development of the community action program, members learned of their shared experiences and dissatisfactions. The au-

15. Fortunately, this incident has only minimal consequence for the longitudinal analysis of organizational effect, since the nonmembers and the newcomers (most of whom had not yet joined the program) were only minimally cognizant of the cut at the time of the interview.

thor of the study, William Zurcher, reports that new members are typically

> surprised that as many members are "in the same boat" and begin to think of themselves as representatives of the poor. . . . Blame begins to be affixed for the suffering, first to a vague and diffuse "them" or "the system" then somewhat more specifically to people with money, assorted community agencies or local government officials. (1969:232–33)

In a study of 400 voters in upstate New York, Wayne Thompson and John Horton discovered that, among the poor and uneducated, the greater one's actual awareness of where the power tends to be centered in their community, the greater is his corresponding increase in negative attitudes toward local school bond proposals. The authors' interpretation is that such awareness leads to frustration, hostility, and, on occasion, open defiance (Thompson and Horton 1962).

Another study of nineteen neighborhood community action clubs in Seattle, Washington, found that when asked to evaluate fourteen different conditions in their neighborhoods, "Club members indicated more dissatisfaction than the nonmembers in every condition, suggesting a greater sensitivity to or awareness of neighborhood problems" (Gove and Costner 1969:650). The members also demonstrated a much greater consensus on the nature of problems facing their communities: no more than 17 per cent of the nonmembers named the same issue, whereas 47 per cent of the members specified housing problems. These findings collectively suggest the presence of a communications network, with an accompanying reality therapy, which supplies members with information concerning their relative status and community issues relevant to their interests—a process which among the poor commonly results in frustration, futility, and anger which may be directed at some political object or the political system in general.

The Enigma of Personal Alienation.   All this does little to explain the apparent tendency of long-term membership in the St. Paul CAA to create, rather than to reduce, *personal* alienation. Nor is there much in the way of hard empirical evidence that would do so. Nevertheless, I suspect that it, too, reflects the communication process available to the poor in community action programs. The explanation is abbreviated in the phrase "ignorance is bliss." It is often said that education raises more questions than it provides answers for and that the truly informed individual is one who has abandoned simplistic descriptions of reality for a more complex understanding of it. To the extent that any practical experience provides an education, along with the enhanced information flow of an organized setting, formerly naïve persons who have found security in ignorance will begin to experience the vagueness and confusion that often accompany awareness.[16] The McClosky and Schaar anomie scale generally measures this type of bewilderment.[17] Similarly, the enculturation process within CAAs may serve to enhance feelings of self-confidence and self-esteem, but the communication process continually impresses upon members their own lowly status and their failure in life by dominant societal standards.

In summary, within any one organization, there may be multiple influences at work, each with the ability to alter existing behavior. There is no reason to assume that their joint impact is cumulative, however. The ultimate effect may be for the various forces to cancel one another. Even if mass society theorists are correct about the propensity of individuals to identify with the collective strength of an association, there may be other organizational influences which serve to negate or actually reverse that effect. In community action agencies, two such forces are the enculturation and communication processes.

THE EFFECT OF PARTICIPATION IN DECISION-MAKING

What of Almond and Verba's contention that the ability to actively participate in the making of organizational decisions leads to a greater sense of personal and political competence? They do not argue that membership alone decreases alienation. In fact, their discussion can be taken to mean that a lack of ability to influence organizational decisions will have adverse effects by increasing one's sense of powerlessness. There is also the question of whether membership alone is capable of having any effect on individual orientations. Roughly half the Head Start respondents interviewed were little more than members in name only, never having donated time at their local centers or attended a PAC or TNC meet-

16. Still another explanation is that mobilization and institutional change, of the type associated with some CAAs, produces a sense of anomie which accompanies the structural anomie described by Émile Durkheim (1951:II, ch. 5).

17. Only five of the original nine items of the scale are used in this study. Items which emphasize nostalgia for the "old days" are omitted. Those which deal with uncertainty are retained.

ing. Only 25 per cent can be said to have been very active in the program over the period of the study. In effect, what this means is that a large proportion of the newcomers were largely untouched by the organization. Their children may have attended the centers on a daily basis, but because the parents either held a job, were taking care of other children at home, or simply did not care, they had little personal contact with the program or the other people in it. It may very well be that the reason the mean difference scores reveal such a slight organizational effect is because newcomers who are members in name only, and therefore unaffected by the socialization process, are averaged in with activists who are affected. Likewise, though there may be forces operating within the CAA which tend to increase political alienation, this does not preclude the possibility that Almond and Verba are correct in assuming that the act of participating in CAA decisions provides still an additional force which tends to reduce alienation.

To explore this possibility, the objective indicator of Head Start political activity, PAC/TNC attendance, was used to determine the effect of involvement on attitude change (table 7.6). For each variable, those Head Start members (both newcomers and old-timers)

TABLE 7.6

RELATIONSHIPS (SOMER'S $\overline{D}$) BETWEEN ATTITUDINAL CHANGE AND POLITICAL PARTICIPATION IN HEAD START †

	Those Who Were Most Alienated at T_1		Those Who Were Least Alienated at T_1
Self-evaluation			
Personal control	.20	*	.02
Self-esteem	.31	**	−.11
Anomie	.0		.15
System evaluation			
Political efficacy	.18	**	−.09
Popular control	.20	**	−.17 *
Government evaluation	−.28	**	−.01
System modifiability	.30	*	.09

† Total membership is 85. The Ns for the dichotomies are: Personal control, 23/57; Self-esteem, 34/46; Anomie, 26/56; Political efficacy, 54/27; Popular control, 43/39; Government evaluation, 26/41; System modifiability, 23/34. Ns for the individual orientation which total less than 85 are a result of missing data from one or both of the two time periods.

* The probability that any randomly chosen pair of subjects is concordant rather than discordant is ≤.05.

** This indicates a $p \leq .01$.

who were interviewed both at T_1 and then again at T_2 ($N = 85$) were divided into two groups: those who were least alienated at T_1 and those who were most alienated. For each of the two groups an analysis was then made of the degree of association between T_2 attitudes and PAC/TNC attendance, thus providing a means of examining the effect of involvement on T_2 attitudes while controlling for original orientations.

A Questionable Confirmation.   The analysis offers only qualified support for the civic culture thesis, because the changes observed also suggest another rival hypothesis: the existence of statistical regression toward the mean. In addition, some members appear to become more alienated as a result of their involvement in the St. Paul CAA. I will proceed to consider the basis for concluding that a positive organizational effect does accompany participation in the St. Paul case and then evaluate the merits of this inference in the light of the rival hypotheses of regression effect and negative effect.

Support for the civic culture thesis rests in the dual facts that activists' orientations at T_2 do differ from attitudes on the same questions at T_1, and that the change is for the most part positively associated with the degree to which members became active in influencing CAA decisions during the interim. Individuals who had quite low opinions of themselves at T_1 and who then became active in Head Start revealed a significant tendency to develop both a greater confidence in themselves and a greater sense of self-worth. Furthermore, the tendency increased the more involved the individual became in PAC/TNC. The same effect is found for political attitudes. Those who were politically alienated at T_1, but who became active in Head Start, were significantly more likely to feel increasingly politically efficacious by T_2 and to believe government officials to be responsive to public wishes in general. Even those activists who continued to express some skepticism revealed a greater faith in the remedial properties of the system than they had been willing to express at T_1. All this suggests that participation in organizational decisions can reduce one's sense of powerlessness as well as create a heightened sense of self-worth.

Yet, in apparent contrast to Almond and Verba's contention that the resulting civic competence leads to system legitimacy, neither PAC/TNC attendance nor the apparent decrease in powerlessness seemed to improve respondents' evaluations of governmental performance. In fact, activists were even more likely than nonactive members to become increasingly dissatisfied with public services. This is not to equate opinions about system *outputs* with appraisals of the *manner* in which decisions leading to outputs are made, the latter being the more common definition of legitimacy. However, dissatisfaction with system performance has as much influence on one's sense of legitimacy as does discontent with the conversion process. If either one becomes too severe, people are likely to lose faith in the political process.[18]

Minimally, the failure of increased satisfaction to accompany a decreased sense of political powerlessness raises questions as to the extent to which changes in one attitude are generalized to changes in related attitudes. If stimulus generalization has occurred in the Head Start case, it is limited to the relationship between personal efficacy and political powerlessness. Even here stimulus generalization appears questionable, given the inconsistent mean changes previously observed between personal control and political efficacy for newcomers and nonmembers as a whole.

In summary, the analysis thus far leads to the conclusion that despite the *enculturation process* peculiar to CAAs, which may intensify political cynicism, and despite the adverse effect of *enhanced communications* in aggravating poor people's feelings of powerlessness, *the opportunity to participate in organizational decisions* which influence one's life and the livelihood of one's family has the stongest impact—one of reducing political alienation. The experience does not create satisfaction with system-wide decisions. Instead, it increases the level of demand. However, it seems to encourage the belief that, if one does not like a decision, she or he, along with others, has some influence in changing them.

Qualifications. There are, however, two important qualifications to such a conclusion. First, it is equally probable that the changes toward reduced alienation are due to statistical regression toward the mean rather than to the opportunity to participate in the organization. The relationship between decreases in alienation and PAC/TNC participation may be simply an artifact of low test-retest correlations for each of the attitudes. Regardless of whether

18. Few Head Start members are so incensed with the performance of governmental officials as to resort to acts of violence, but those who are dissatisfied are significantly more likely to engage in moderate acts of dissension. The relationship (Tau $b = -.28$) is significant at the .001 level. "Moderate dissent" here consists of verbal criticism of public officials, both through the mail and in the presence of friends and neighbors.

imperfect test-retest correlations are due to error or to systematic sources of variation, their existence creates the likelihood of falsely inferring causality in what is actually a case of random change. The fallacy stems from the tendency of individuals *selected because of their extremity* on scores at T_1 to regress toward the T_2 mean of the population from which they were selected (Campbell and Stanley 1963:10–12). Campbell observes that "Regression artifacts are probably the most recurrent form of self-deception in the experimental social reform literature" and that it is particularly common in quasi-experiments conducted in natural settings where self-selection to treatment groups occurs (1969: 413–14).

This is the case with the St. Paul Head Start program. Members who chose to become active participants in the organization were more extreme in their attitudes at T_1 than those who did not. At least a portion of the association between attitude change and PAC/TNC attendance reflects that fact, and not organizational effect. This is a difficult but vital point. Further clarification is in order. Normally the regression fallacy would not apply to the type of analysis conducted here because I did not simply compare the means or percentages at T_1 with those at T_2. Instead, changes over time from extreme attitudes at T_1 are *correlated* with level of activity during the interim. The fallacy therefore could be said not to apply *if* there is no interaction between attitudes at T_1 and PAC/TNC attendance. However, there is such an interaction.

As previously observed at T_1, newcomers as a whole were generally less personally and politically alienated than nonmembers. Yet *within* the organization, those who became more active in the internal politics of the program (PAC/TNC meetings) were *more* politically, as well as personally, alienated than the less active or apathetic members.[19] The only exception is on the question of system reform, where activists were more optimistic than nonactivists. Thus the very people who were most likely to participate in PAC/TNC meetings are the very same people most likely to regress by chance toward a lower level of alienation by T_2. In this case it therefore cannot be determined whether the reduction in alienation among activists is due to an organizational effect or a

19. The reason that alienation of either type leads to greater involvement within the organization stems from the peculiar nature of CAAs and the groups to which they most appeal. In the light of other evidence from national cross-sectional samples which indicates that normally it is the least personally and politically alienated who participate politically (Campbell et al. 1960:103–4, 518–19), this subject is of interest in itself and will accordingly be discussed in a future article.

regression effect. It is most certainly due, at least in part, to the latter.

There is another, though weaker, tendency among Head Start activists: one toward greater personal and political alienation. Unlike the change toward less alienation discussed above, this one is *concealed* by the regression artifact rather than produced by it. Specifically, some members who at T_1 were relatively unalienated appear to have become more personally and politically alienated at T_2 as a result of their activity in CAA. Only three of the associations approach statistical significance. Activists are inclined to develop a lower sense of self-esteem ($p = .08$), a higher sense of anomie ($p = .10$), and a greater skepticism about the responsiveness of public officials to popular wishes ($p = .05$). These findings are consistent with results examined thus far in which simple membership and membership of long duration are both associated with heightened alienation.

The conclusion that these tendencies are concealed by the regression effect stems from the nature of self-selection in this particular case. People who were chosen in the analysis for their extremely high scores at T_1 are just as likely to regress toward the group mean at T_2 as those who are chosen for their extreme low scores. However, since the activists are the most alienated of the Head Start group, the interaction between attitudes at T_1 and PAC/TNC attendance does not reflect statistical regression as it did in the preceding analysis. It runs counter to it. Thus, if anything, the tendency of some members to become more alienated as a result of organized activity is partially obscured by the other tendency of activists to regress from negative scores at T_1 toward the means at T_2.

CONCLUSION

The purpose of experimental and quasi-experimental research is to prevent the researcher from merely computing evidence in favor of a preferred hypothesis. It provides a harsh but fair test of his propositions. Under such scrutiny neither the mass society nor the civic culture theory fares very well. After examining them from three perspectives—simple membership, membership of long duration, and active membership—there is little basis for accepting their validity. Only in one analysis, that of active participation in Head Start decisions, is there any evidence that CAAs can reduce personal or political alienation. However, at least part, if not all, of this relationship is a statistical artifact.

Over-all, the St. Paul Head Start program has little apparent impact on members' psychological orientations. What effect it does have aggravates rather than reduces alienation. Differences in mean attitude changes over time between members and non-members approach statistical significance only in a few cases: personal control, self-esteem, and anomie for old-timers, and anomie alone for newcomers. Each of these changes is toward greater personal alienation. Similarly, those changes associated with degree of organizational involvement, which are free of the alternative statistical interpretation, are all negative. Formerly alienated activists become even more dissatisfied with governmental performance, and formerly nonalienated activists demonstrate a propensity toward a lower self-esteem, heightened anomie, and a greater skepticism about the responsiveness of public officials. The only bright side of the picture is that membership of long duration and active involvement show slight signs of moderating political alienation by maintaining or restoring a faith in the possibility of system reform.

The implications of these findings, particularly if they are substantiated by subsequent research, are both depressing and stimulating. They are depressing because they indicate the failure of community action to restore to the poor a sense of confidence and satisfaction regarding one's own self and the community in which he or she resides. They are stimulating because they open the door to a fuller understanding of the diverse effects organized activity may have on psychological orientations.

I have previously noted that the St. Paul Head Start program offers optimum conditions for a confirmation of the mass society and civic culture theses. Yet it does not provide that confirmation. Further research will presumably uncover the reasons for its failure to do so. It may be because there is no merit to either thesis. On the other hand, scholars like Lipset et al., Seeman, Almond and Verba may be correct, but only under certain circumstances. I have already suggested that three of these circumstances involve the nature of the membership, the communications, and the enculturation process peculiar to an organization. I will conclude by suggesting three additional conditions which probably affect the ability of organizational involvement to alter attitudes. Since the effect of Head Start was to aggravate alienation rather than reduce it, I will posit only those conditions which are likely to contribute to the same outcome. It is assumed that members do participate in the organizational decision-making process.[20]

First, it may be that, though members can influence decisions, they are nevertheless dissatisfied with their ultimate outcome. The civic culture thesis fails to consider organizational goal attainment and its interaction with participation in the making of decisions. Perhaps participation is only likely to reduce alienation in those cases where one is satisfied with decisions. Feeling that one's actions have benefited his or her own life may lead to heightened self-confidence and self-worth. However, the belief that one's actions have helped to produce a situation detrimental to one's interest can be expected to have just the opposite impact.

Second, the act of participation in organizational decisions does not necessarily entail the belief that in fact one is influential in that process. Some members may have simply concluded that, despite their effort, they personally—or perhaps members in general—have little real influence in the program. The participation, in other words, may not be meaningful. The actions of parents in PAC/TNC may be little more than a formality, designed to create a false sense of satisfaction or legitimacy—or at least some parents may perceive this to be the case.

Finally, it is quite possible that some participants simply expect too much from the agency. They receive indoctrination which emphasizes the advantage of community action over previous types of antipoverty programs of a more degrading and less responsive nature. It is quite possible that a few members overreact to this stimulus, develop unrealistic, overinflated expectations of the organization (similar to the phenomenon of rising expectations common in modernizing nations), and, after being disappointed by the practical limits to the program, become frustrated, suffering adverse effects.

Whatever the explanation, at least two points are clear. First, organizational membership alone has only limited ability to influence attitudes. Some degree of involvement is necessary. Second, the effect which does occur is not necessarily one of decreased personal and political alienation. There are multiple forces at work within any organizational context, some of which alleviate alienation while others aggravate it. To the extent that these forces have a congruent impact, the socialization process will be easy to discern and predict. To the extent that they are incongruent, organizational effect will vary depending on the peculiarities of the organization and of the individual participants.

20. I plan to present, in the near future, an empirical discussion of these possibilities based on the St. Paul study.

8

COMPLIANCE, POLLUTION, AND EVALUATION
A Research Design

LESLIE L. ROOS, JR.

HAL J. BOHNER

Starting at the most general level, a basic model appropriate for water management research (Mar 1971) is:

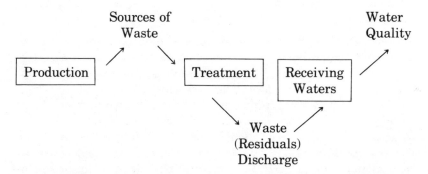

Such a model suggests several ways in which data relevant for the evaluation of various enforcement strategies could be collected. If we look at industrial pollution at the factory level, changes in process recycling, treatment facilities, and residuals discharge would be useful for evaluating the efficacy of any strategy.

Previously reported quasi-experimental research on pollution (Caldwell with Roos 1971) emphasized water quality data, the last link in the chain presented above. In lieu of direct measurement of waste discharge, it would be desirable to measure changes in water quality so as to permit more precise inference about change in residuals discharge by particular sources.

This earlier research (Caldwell with Roos 1971) used an inter-rupted time-series design to investigate pollution control efforts in the lower Lake Michigan area during the 1965–69 period.[1] The "event" used to divide the time series into "before" and "after" segments was a negotiated compliance deadline (December 31, 1968) by which time several oil and steel industries were to have made significant improvements in their effluent discharges. Various chemical parameter values prior to the end of 1968 and aggregated into six-month mean scores were viewed as pretest scores; scores subsequent to that date were considered as post-test.[2]

In this pilot study, data were available from three sampling sites near three large steel companies' plants (U.S. Steel, Inland Steel, and Youngstown Sheet and Tube) and three large oil companies' refineries (Mobil Oil, Cities Service, and Sinclair); one control site was used, helping to eliminate several plausible rival explana-tions. Information was gathered for six different pollutants. Of the 24 trends (4 sites × 6 pollutants) tested, the following postcompli-ance tally resulted:

Upward	Downward	Irregular	Level
5	3	8	8

These results called into question the efficacy of depending on voluntary compliance schedules, but, because only two "after" time points were available, additional data from more recent time points would raise one's confidence in the findings.

Certain conceptual problems with this research need to be discussed; a distinction between policy outputs and policy outcomes (Jacob and Lipsky 1968) should be made. Outputs are such things as the passage of laws and enforcement efforts; outcomes are effects upon water quality, quality of the air, and so forth. The output of business-government interaction may not result in better outcomes. The analysis of compliance may have more general implications. The problem concerns factories which may have satisfied the various regulations or wishes of enforcement agencies. Such compliance may, in fact, have been ineffective in reducing

1. The interrupted time-series design is discussed in several essays in this volume. Some problems of measurement error and autocorrelated residuals are treated in the essays by Roos and by Duvall and Welfling, respectively.

2. Aggregation of the data semiannually greatly reduces the number of obser-vations and does not take into account variance within each six-month period. A number of statistical problems are raised by this use of mean scores. These data might profitably be reanalyzed with little or no aggregation.

effluent discharged from that same factory. The output (meeting standards or regulations) may be quite different from the outcome (polluted water). This problem can be rephrased from an engineering perspective. As Mar (1971) has pointed out, treatment criteria (output criteria) are not the same as water quality criteria (outcome criteria). The pollution problem might well be studied at each stage in the process.

At this point, it is useful to review the several assumptions which characterized the previously reported study of pollution in the Lake Michigan area:

1. Water quality data can be used to make inferences about residual discharge (waste) from factories.
2. Water quality would not have become worse without the negotiated compliance program.
3. Any changes in factory residuals discharge stimulated by the compliance program would have been picked up in the one-year period after the deadline passed.

All these assumptions should be discussed more fully.

PROBLEMS OF MEASUREMENT

Using water quality data to make inferences about residuals discharge from factories poses a number of problems. Water quality measures from both the downstream sampling point and the upstream sampling point nearest the outfall of a given plant would be valuable for these purposes. This "dual" data collection would help isolate the effects of the particular plant on water quality. Ideally one would want:

1. Site 1 immediately upstream of the effluent outfall. Data from this site would give a measure of water quality before the contribution of the effluent from the particular outfall.
2. Site 2 immediately downstream of the effluent outfall. Data would provide a measure of water quality upstream from the site (using a given indicator) *plus* contribution of the effluent from the particular outfall.
3. Streamflow data corresponding to the date on which the upstream and downstream samples were taken. Flow, expressed in cubic feet/second, is a standard measure.

By subtracting the concentration for Site 1 from that for Site 2, and correcting for dilution using the streamflow data, it should be possible to isolate the contribution of the effluent from a particular outfall. With Site 1 and Site 2 very close together, the problem of various stream processes affecting the effluent would not arise. In reality, our ability to interpret such data depends on there being comparatively few major outfalls within a given stretch of stream. Thus, some plants may defy evaluation because of the presence of too many outfalls per site.

The importance of various stream processes will vary according to the indicator involved. Indicators based on dissolved oxygen might change markedly if a sizable stretch of stream is between the upstream and downstream sites; reaeration, benthic load, and a number of other parameters particular to a specific stream would be critical. Conservative indicators of water quality — such as sulfate and cyanide — would be less susceptible to such changes; therefore, nondegradable pollutants should be studied wherever appropriate. The procedure might be modified in the case of pollutants for which only qualitative data is recorded by water quality surveillance agencies, e.g., presence and/or absence of floating oil.

Preliminary research has been conducted using the methodology suggested here. Such parameters as specific conductivity, sulfate, alkalinity, and nitrate were found to be predictable at intermediate sites (i.e., estimates could be made about effluence discharge) on the basis of upstream and downstream measurements. Predicted and actual data were characterized by high correlation coefficients (.91 to .99); the corresponding regression coefficients had narrow confidence intervals. Such findings strongly suggest that, for these findings, the relationships between residuals discharge and water quality can be established. Other investigators have also made progress on this problem (Mar 1971). These parameters which appear predictable are good candidates for further investigation and use in an inferential model.

Variations in the persistence of different indicators suggest one way in which it might be possible to separate out the effects due to several outfalls in one stretch of river. Certain types of indicators can be generally associated with specific production processes. For example, the iron-steel industry produces a very different mix of effluents than does the oil industry. If one outfall from each type of plant (one from iron-steel and one from oil) were located between an upstream and a downstream sampling point, valid inferences might well be made by concentrating upon the indicators relating to materials produced by each plant. In many industrial discharges a correlation among the various parameters through time may be found. Measures of the correlation among such parameters could lead to a method whereby industries discharging the same pollutant in a study area can be analyzed.

In the quasi-experimental study of lower Lake Michigan, only downstream sampling sites were used. Thus, the possibility of changes in upstream residuals discharge poses a threat to the validity of the inferences about the oil and iron-steel factories and should have been treated more specifically in previous research (Caldwell with Roos 1971).

EFFLUENT DISCHARGE, PRODUCTION, AND CAUSAL LAGS

The second alternate hypothesis focuses on the relationship between volume of production and effluent discharge: the effluent discharge (and hence water quality) would have been substantially worse without the companies' efforts spurred by the voluntary compliance agreement. Why might this have been so? If production has been markedly higher in 1969 (after the compliance deadline), then in the normal course of events residual discharges would have been greater. The only figures available on steel production were national in coverage, showing that ingot steel production for the three leading types of furnaces increased 6–7 per cent between 1968 and 1969 (Institute of Scrap Iron and Steel, Inc. 1970:37). This provides some support for the idea that the voluntary compliance program was not successful; the likely increase in production was not so great as to lead to substantially increased pollution.

A more serious problem concerns time lags. One chain of events might be:

State Enforcement Effort	Negotiated Compliance Agreement	Technical Effort	Installation of Equipment	Change in Residuals Discharge
→	→	→	→	

Several years might pass before there was any change in residuals discharge. The passage of more time after the compliance agreement would have been useful in judging the strategy's effectiveness. In a personal communication to the authors, a steel company representative said that these firms were too optimistic in thinking that changes could be made by the original compliance deadline, but that progress was being made by a year or so after the deadline. The nature of the change — if any — instigated by enforcement efforts should be noted. If the change occurred in such a delayed fashion it would not have been picked up by the interrupted time-series techniques. Additional data collection and analysis would certainly be appropriate here.

Finally, the impact and goals of a public program should be

considered in terms of the total effort being made in a specific area. If the "voluntary" compliance program had been used as a small portion of an existing over-all strategy, short-term failure would not be important. But when a program constitutes a major part of over-all strategy – as did the negotiated agreements in the late 1960s – a lack of isolatable short-term effects becomes important. The negotiated compliance agreements may have had longer-term effects; although such effects are likely to be difficult to isolate, the research design forwarded later in this paper will attempt to detect them.

AN EXPANDED DESIGN

The generalizability of the findings from small-scale studies is always subject to question. A larger project might evaluate both more industries and more strategies than did the Caldwell with Roos (1971) study; such a project should allow for a more powerful design than that used in the cross-sectional research described above. The pulp-paper industry might profitably be considered along with iron-steel and oil. These three industries contribute a substantial proportion of the total industrial residuals deposited in the state's waterways. Pulp-paper, iron-steel, and oil were the three industries most frequently in attendance at Interstate Enforcement Conferences.

Given the dependent variables (various pollutants), four major strategies might be evaluated on a factory-by-factory basis:
1. Enforcement conferences and their specific emphasis on voluntary compliance;
2. Legal sanctions against specific companies and plants;
3. General inspection and state-level enforcement efforts not involving litigation; and
4. The federal permit program. Recent court decisions have made this program difficult to evaluate, but the situation could change at any time.

The first two sorts of enforcement effort have been directed toward some companies and plants, but not all. Although the timing of the enforcement effort also varies from plant to plant, use of a particular strategy should be abrupt enough to divide the data into before and after components. The permit program will be affecting all the different plants at more or less the same time; here again, a plant's receiving a permit is an unambiguous event which should allow analysis on a before-after basis. Thus, the interrupted time-series design seems generally feasible.

Some broader features of the research design should be treated. Policy research must be concerned with the matching of variance in the independent variable with appropriate occurrence of the specific phenomena (dependent variables) under study. Enforcement strategy is the independent variable, but operationalization is necessary. For this operationalization, it is important to specify which state employs which strategy. States could be selected with an eye to including a wide range of pollution control strategies.

Differences among the states exist along several dimensions. The effects of these variables upon water quality is not known; these differences make it possible to think in terms of "most different systems" designs and multiple replications. Wenner (1972) has ranked the six states along these criteria; although the rankings in table 8.1 are somewhat dated, they are still useful.

TABLE 8.1
ENFORCEMENT EFFORT, LEGISLATIVE INTEREST,
AND RESOURCES DEVOTED BY STATE

State	Enforcement Effort	Legis- lative Interest	Resources Devoted to Water Pollution Control
Illinois	1 (out of 50)	18	30
Ohio	4	2	36
Wisconsin	17	6	1
Indiana	18	31	20
Alabama	33	38	50
Washington	36	20	3

In addition to these quantitative differences, there are qualitative differences among state approaches to environmental protection. Ohio's effort has been characterized by extensive planning and coordination for the Ohio River area. Such efforts at coordination have been heavily "promoted" by professionals in the environmental protection area. Illinois, for example, has had unusually extensive involvement with legal and quasi-legal proceedings through both the state Environmental Protection Agency and the state attorney general's office. At the state level, there would be some problem in separating the amount of enforcement effort from the type of enforcement effort. This problem might be partially resolved by looking at the quantity and type of influence which have been exerted at the factory level: visits, voluntary compliance, threat of litigation, etc. Amount of enforcement effort would presumably be associated with "quantity"; states similar in amount of enforcement might differ as to "type."

Industries are not distributed randomly across geographic areas. Their distribution presents some problems in evaluation; it is difficult to simultaneously consider both state enforcement strategy and type of industry involved. As is seen in table 8.2, certain industries are concentrated in particular states.

TABLE 8.2
APPROXIMATE DISTRIBUTION OF INDUSTRIES BY STATE

State	Iron-Steel	Oil	Pulp-Paper
Illinois	Much	Little	Little
Ohio	Much	Some	Some
Wisconsin	None	None	Much
Indiana	Much	Much	None
Alabama	Much	None	Much
Washington	None	Some	Much

An overlap among types of industry and type of enforcement strategy is necessary to evaluate the generalizability of a particular strategy which may have been efficacious in one set of circumstances. Such generalizability does pose design problems; there is sufficient overlap among industries and states to permit some comparisons which do control for industry.

In each of the three industries some firms and plants have been minimally involved with enforcement conferences and legal actions. Variation is necessary for this design, since such differences among firms would define the "treatment groups." Although distinguishing between different kinds of legal sanctions may be a necessity, the basic logic should be clear. Comparing firms and plants with very different enforcement histories provides several sorts of control groups. Such comparison groups — particularly those plants which have been subject to minimal enforcement action — would help determine whether trends independent of specific enforcement efforts have played a major role in changes in water quality.

The cumulative nature of pollution abatement policies does pose some problems. Enforcement conferences and voluntary compliance may have no effect upon a given company at one point in time, but, in combination with other policies, may be influential at a later date. This is a particular difficulty because of the sequential nature of many of the enforcement efforts. There are two quite different approaches to dealing with these problems of interaction:

1. trying to find enough instances of particular policies occurring both independently of each other and in combination with each other to permit accurate estimation; and
2. concentrating upon main effects and—if there is no main effect—assuming the particular variable under consideration to be unimportant in the multivariable case.

It would clearly be most desirable to have enough cases to use the first approach. The number of factories in the six states under consideration should permit application of the first approach.

Cross-sectional research reported elsewhere (Roos and Roos 1972) has indicated some differences among companies and among factories in efforts to reduce pollution. Although relationships were rather weak, larger companies and larger factories tended to do more in terms of effluent treatment than do smaller operations, even though these efforts do not seem to be reflected in terms of such effluent measures as BOD discharge. This design permits studying such differences among companies and among factories in considerably more detail than has been possible previously.

Specific propositions have been generated from research cited earlier. It is hypothesized that:

1. Enforcement conferences and voluntary compliance will have proved ineffective in many cases.
2. Legal actions against particular companies and particular plants will have proved to be somewhat effective.
3. There will be substantial variation among industries and among companies in their response to enforcement efforts.
4. There will be substantial variation among states in the effectiveness of their enforcement efforts.
5. Relatively little experience is available for the permit program. We suggest that the effectiveness of the permit program will depend on the specific enforcement efforts keyed to its application in each state. Questions as to when —and if—permits will start to be issued complicate matters here.

It should be noted that this design focuses on programs and enforcement efforts; other commentators on environmental policy have emphasized institutions and institution-building (Craine 1971). Although I believe a focus on enforcement to be potentially fruitful, an interrupted time-series design with multiple replications would seem equally applicable to the evaluation of institutions. The strategy proposed here would help alleviate problems intrinsic to the research discussed above (Caldwell with Roos 1971; Roos and Roos 1972). The collection of both comparative and longitudinal data would aid in eliminating plausible hypotheses which did not stand up under the various sorts of statistical tests. The goal is to specify more adequately the particular conditions under which one sort of enforcement strategy appears to work better than another.

9

EVALUATION, QUASI-EXPERIMENTATION, AND PUBLIC POLICY

NORALOU P. ROOS

Congress has directed federal agencies to set aside one per cent of their funds to be used explicitly for program evaluation. This has created a tremendous reserve of money for evaluation research, and one might assume rapid progress toward what Campbell (chapter 6) has called "an experimental approach to social reform." However, most appraisals of evaluation efforts are discouraging (Wholey et al. 1970). The need for project evaluation and the development of output measures is still being discussed. Relatively few evaluation efforts have had a subsequent input into the policy-making process.

Many explanations have been advanced for the neglect of evaluation research. Weiss and Rein (1970) have argued that experimental evaluation of programs with broad aims is so fraught with difficulties that it should not even be attempted. They believe such research is neither valid nor useful to the policy maker or administrator. Campbell (1970) responds that Weiss and Rein's alternative, process research, might usefully supplement experimental evaluation research, but should not replace it. Campbell continues to seek the development of "methods for the experimenting society" (1972). His work and that of others represented in this book is based on the hope that, if we are able to do valid evaluation research, it will be used.

This paper was written while the author was working in the National Center for Health Services Research and Development, DHEW, as a part of the AACSB-Sears Faculty Fellow Program. The opinions expressed are those of the author only.

281

Wholey et al.'s (1970) analysis of the status of evaluation in fifteen programs conducted by four federal agencies raises doubts about even this mildly optimistic prediction. The Office of Economic Opportunity was found to be the agency conducting the most extensive evaluation activity. However, even in OEO very few examples of evaluation research having a subsequent impact on policy or operations were found. Four basic reasons for the low utilization are cited: organizational inertia, methodological weakness, design irrelevance, and lack of dissemination. Since these authors feel existing methodology is adequate for the evaluation of most social programs, their study suggests that the major barriers to conducting valid and useful evaluation research are organizational, political, and behavioral. Thorner's (1971) analysis of the lack of input of evaluation research into health policy planning concurs with this view.

Each of these explanations captures part of the picture. However, none of them systematically attacks the problems confronting the successful implementation and utilization of evaluation research. Coleman (1971) has made a start in this direction by calling for the development of a coherent, self-conscious methodology for studying the impact of public policy. The thrust of Coleman's argument can be summarized by stating that there are two important requirements for evaluation research: (1) the research must be valid, and (2) the research must be useful to policy makers or administrators.

Researchers consider validity to be of obvious importance; most of the articles in this book are directed at improving the scientific credibility of evaluation research. Usefulness is equally important to the policy maker. Usefulness is related to validity but must encompass such intangibles as the type of information the policy maker is willing to use and the ability of the policy maker to get results in time to use them.

Researchers argue that, if the evaluation is not valid, it will be worse than useless to a policy maker—it may lead him astray. As strong an argument can be made about the need for usefulness; if evaluation efforts are ignored by the decision maker, their validity is somewhat beside the point.

Unfortunately, validity and usefulness are probably incompatible objectives. The time and resources necessary to obtain valid results acceptable to researchers often make such efforts useless to policy makers. Similarly, policy makers by design or default frequently confuse public relations with evaluation research. In their quest for meeting immediate policy needs they are content with superficial evaluations which beg validity questions. Figure 9.1 presents the typical positions of the policy maker and the researcher. This paper argues that evaluation research must search out the middle ground between the needs and concerns of these two groups.

FIGURE 9.1.   Existing Researcher and Policy Maker Conceptions of Evaluation Research, and a Proposed Alternative

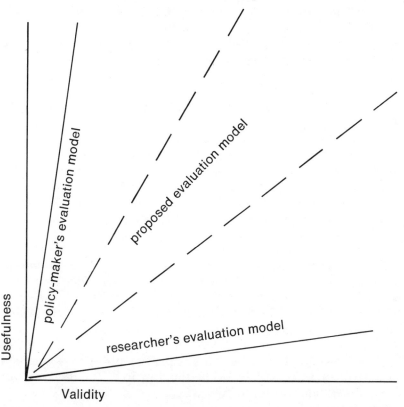

It is proposed that researchers must discover the range of validity which will provide meaningful guidance to the policy maker. The more time and resources available, the more effort which can be devoted to raising confidence in the results of evaluation. However, when time and resources are in short supply, both the policy maker and the researcher need guidelines for doing and using evaluations.

This paper attempts to develop a general framework for examining threats to the validity and utility of evaluation research. The evaluation research model is first presented, and then four different types of threats to evaluation efforts are reviewed: (1) threats due to the nature of the policy maker's objectives, (2) threats due to the inability to adequately cover program alternatives, (3) threats due to difficulties of assessing program impact, and (4) threats due to difficulties of implementing research results.

The discussion draws upon familiar theoretical and disciplinary traditions for developing strategies to cope with these threats. This examination also points up the interrelatedness of evaluation research; what occurs at one stage can greatly impact subsequent stages. In a final section of the paper, guidelines for conducting evaluation research are proposed.

THE EVALUATION RESEARCH MODEL

The evaluation research model (figure 9.2) focuses on systematically relating program inputs to program outcomes or impacts. This model assumes that the policy maker has an objective, such as "more efficient delivery of health care," which he wishes to achieve. Of course, he may have several objectives in changing the delivery of health care: improved efficiency, improved quality, and improved access, for instance. A second part of the evaluation model is a consideration of alternative means for achieving these objectives. The alternatives examined may be only implementation of Program 1 versus lack of implementation. The implementation of several alternative programs could also be examined. The third element of the model is impact analysis. This focuses upon which program or combination of programs delivers the most in terms of the objective(s). The final part of the model is implementation of research results. It assumes that the policy maker or the administrator will be influenced by evaluation research. The research will help him decide which programs to fund or curtail, which laws to pass, and so forth.

FIGURE 9.2.   The Input-Impact Evaluation Research Model

Policy Maker's Objective	Input via Program Alternatives	Assess Program Impact	Implementation of Research Results
A	1	more A	decision to fund program 1
	2	less A	

THREATS TO EVALUATION RESEARCH

In certain important respects, the evaluation research model resembles the rational decision-making model used by economists. Both models focus upon the decision maker's objectives, the identification of important alternatives, and the necessity of relating the alternatives to anticipated outcomes. Given these conditions, the decision-making or policy process becomes the rather obvious one of choosing the alternative which achieves most of the desired objectives.

The major difference between the evaluation model and that used by economists is the time framework. The economists' model generally assumes the *existence* of information relating alternatives to outcomes, while the major concern of evaluation research has been the *acquiring* of information necessary to relate outcomes to alternatives. However, the time question is relative; once the information is available, the models are very similar. This means that difficulties found in applying the economists' model to real world decisions may also be found in implementing the evaluation model. Moreover, just as the appropriateness of decision processes varies according to the characteristics of the decision situation (Mack 1971), so may the appropriateness of a given type of evaluation vary, depending upon the nature of the program to be evaluated.

Previously posed explanations for the minimal impact of evaluation research have had an ad hoc flavor. By separating the threats to such research according to the element of the model affected, a better understanding of the problems can be gained.

THREATS DUE TO THE NATURE OF THE POLICY MAKER'S OBJECTIVES

Evaluation researchers have long recognized the importance of identifying program objectives. Wholey et al. (1970) list this as the first requirement of an evaluation system, while at the same time noting the difficulty inherent in this task. They stress that most federal programs have multiple objectives, and that there is at best a poor understanding of the relationship of program activities to ultimate agency goals. Even when a single policy objective is agreed on, that same objective can mean quite different things to different people.

Researchers have addressed this issue. Eckenrode (1965) has described different techniques for clarifying goal structures, and others (Golembiewski and Blumberg 1967; Huber and Delbecq

1972) have experimented with techniques for moving toward goal agreement. No real solution, however, has been found.

Evaluation researchers do not agree about what should be done when goal diversity cannot be reduced. Weiss and Rein (1970) argue that a program with broad aims and multiple objectives is inherently unsuitable for experimental evaluation. They suggest as an alternative *process evaluation.* By this they mean an evaluation which collects qualitative as well as quantitative data, and which focuses on a complete description of events as they develop through time. A case study approach to evaluation research, rather than a hypothesis testing approach, is advocated. Campbell (1970) agrees that some programs deny meaningful experimental evaluation, but he believes that no evaluation at all is better than using descriptive activities and then calling them an evaluation. He argues that multiple modes of measurement can be used to tap several goals.

As plausible as this may sound, quasi-experimental evaluation research has typically focused on few goals, and sometimes on only one. For example, Roos and Bohner have proposed in chapter 8 to study the impact of regulations upon water quality. Multiple indicators [1] of water quality are suggested, but the entire focus is on this one objective.

As important as water quality may be, such an evaluation will not touch other important objectives of the policy maker, for example, cost of such enforcement procedures to the companies and the consumers, and the impact of such enforcement upon plant closings and subsequent unemployment. Campbell (chapter 6) has discussed several impacts produced by the Connecticut crackdown on speeding. He examines the impact of the crackdown on traffic fatalities, on speeding (indirectly, by comparing the proportion of licenses suspended for speeding with those suspended for other reasons), on the reaction of the courts (proportion of speeding violations judged nonguilty), and on reactions of individuals to harsh punishment (proportion arrested while driving with a suspended license). Even this imaginative evaluation fails to cover many objectives important to the policy maker. What impact did the program have on police expenditures? If no great additional expenditures occurred, were police diverted from other

---

1. Dealing with multiple indicators of one goal should not be confused with dealing with multiple goals. The former is more prevalent in quasi-experimental evaluation research. Thus Campbell (chapter 6) has proposed to examine posttraining employment of Job Corps trainees using two indicators: (1) amount of withholding tax withheld two years later, and (2) percentage drawing unemployment insurance.

important activities? What was public response to the program? Was there a backlash, or general acceptance and support? Monitoring these additional objectives would have been especially important if the evaluation had concluded that the program had a strong impact on reducing traffic fatalities. Another administrator attracted to the crackdown approach for reducing traffic fatalities would want to know what it would cost, and how the public would react.

The evaluation researcher could conceivably develop indicators for as many goals as the policy maker has. However, the cost of such activity would undoubtedly be high, particularly when data collection must begin from scratch.

Advocates of quasi-experimental evaluation research are only beginning to address the question of aggregating the results obtained in a multigoal evaluation (Campbell 1972). Will the policy maker simply be told that Policy 1 has a positive effect upon *A,* no effect upon *B,* and a negative effect upon *C?* How will the strength of impacts be compared when there is no common measuring stick? The decision-making literature has anticipated much of this discussion. Economists working on problems of cost-benefit analysis have devoted considerable thought to problems of evaluating noncommensurate objectives (Freeman 1970).

THREATS DUE TO INABILITY TO ADEQUATELY COVER
PROGRAM ALTERNATIVES

Evaluation researchers are somewhat ambiguous in their treatment of program alternatives. Much of the research has focused on a single program and has attempted to define the effect, if any, of the program.

Wholey et al. (1970) found this single program focus throughout federal evaluation activities: ad hoc evaluations are commissioned with little possibility for making comparisons across programs. This single program focus is also true of the quasi-experimental evaluations discussed by Campbell (chapter 6). In the Connecticut crackdown example, only one alternative method for reducing highway fatalities was examined. An administrator who wants to use such results for planning his own program would find it much more useful if several ways of reducing traffic fatalities had been compared: How does a speeding crackdown compare with a publicity campaign, or a law requiring the installation of advanced safety features?

Evaluation researchers, of course, do not advocate a single

program focus. Cook and Scioli (1972) outline a factorial evaluation design which policy makers could use for comparing programs
according to their ability to reduce different types of pollution
coming from different sources. However, the costs of exhaustively
evaluating program alternatives in time, effort, and monetary
expenditures would certainly be high.

Recognizing this problem, Wholey et al. (1970) have recommended that evaluation funds be pooled. They advocate eliminating the single project evaluation in favor of large evaluation
studies and using the same methodology to compare what happens
in several locations. Another approach would be to develop standardized, or at least compatible, data systems. There is a shocking
lack of such compatibility in federal, state, and local data collection efforts. Compatible data systems would make comparisons
across discrete evaluations more feasible. This, combined with a
move to more standardized data collection instruments and to
data archives making past evaluations generally accessible,
would greatly increase the possibilities for the analysis of policy
alternatives. Moreover, data archives would increase the possibilities for designing new quasi-experiments built upon a combination of old and new data (Hyman 1972; Roos and Roos 1971).

Unfortunately, the data problems are not the only ones which
must be solved if multiple alternatives are to be successfully
evaluated. Major methodological efforts may also be necessary. It
can be argued that some of the attractive features of the interrupted time-series analysis will be lost if such analysis is directed
toward the comparative evaluation of two or more program alternatives.[2] For example, it would seem easier to detect a jerky
movement against a stable environment than to assess differences
between two jerky movements (programs). This is especially true
given the lack of a measure of strength of association within one
interrupted time series (see chapter 1). Without this measure,
program alternatives evaluated by an interrupted time-series
design could only be compared by visual inspection of intercept
change or slope change, or by comparing tests of significance.
Interrupted time-series analysis also relies on successive observations occurring in a similar content universe. Comparative evaluation of two or more program alternatives will tend to erase the
advantages of this similar content universe, since the alternative
programs will of necessity be administered simultaneously in

2. The author is indebted to James Caporaso for the observations in this para-
graph.

different environments or consecutively in the "same" environment.

The difficulties inherent in evaluating several alternatives may limit the usefulness of evaluation research as an input to the planning process. This issue will be discussed further in the final section of this chapter.

THREATS DUE TO DIFFICULTIES OF ASSESSING PROGRAM IMPACT

Articles in this book address themselves to methods appropriate for improving the internal and external validity of evaluation research. The following steps have been recommended: (1) use of multiple indicators; (2) sampling of an increased number of time points; (3) performance of advanced data analysis; and (4) use of an advanced research design. Unfortunately, as has also been observed by Caporaso in chapter 1, a theoretical framework for guiding the application of these methods has not yet been adequately developed. The evaluation researcher needs some rationale for rank ordering these methods, or for knowing which method or combination of methods is appropriate for a given research situation. The researcher also needs guidance in deciding how many indicators or how many time points to use. Roos says in chapter 2 that several time points need to be measured, but what are the costs and benefits associated with gathering data from more points in time? At what time intervals should the data be collected? Finally, and most important, when are the results of an evaluation *valid enough* to guide policy making? Endless methodological refinements are undoubtedly possible, but when is it appropriate to stop designing, measuring, and evaluating and to start recommending?

Roos has also given some ad hoc advice on when to use one method and when to use several. Campbell (chapter 6) distinguishes designs according to how much administrative foresight is required for their design and implementation. Campbell (1970) has also briefly indicated how one might systematically improve descriptive process research and move it toward quasi-experimental and experimental evaluation. He emphasizes the need for gathering data prior to the intervention and for establishing a control group.

More attention needs to be paid to this methodologically unsophisticated end of the evaluation spectrum. Most activity carrying the label of evaluation in government today is being done in the process tradition rather than in the quasi-experimental tradition. Descriptive or process research is used to obtain information on the general political feasibility of a program and to provide gross effectiveness data: Can quality assurance or physician peer review function at all? At this level the criterion being used to evaluate a program is "Does it survive?" Process research is also used by policy makers to draw general guidelines about how to replicate a trial program and to identify the legal, organizational, and practical problems which must be overcome. However, process analysis is not input-impact evaluation in the sense that we have been using the word.

Process analysis typically documents what happens once a change has been introduced. Little attention is paid to preintervention events, or to what is happening in organizations where the intervention is not being made. Since political factors and quality of proposal, rather than planned variations, usually determine the sites within which interventions are made, it becomes difficult if not impossible to determine why program A had impact B. Thus, unless change per se is the policy maker's objective, process analysis usually does not provide a systematic basis for deciding whether or not a trial program should be more widely implemented. Unfortunately, this distinction between *how* to replicate a program and *whether* a program should be expanded is not often clearly made.

This does not mean that process research is useless. Process research is analogous to a preliminary case study. Where there is little understanding about basic relationships involved, a descriptive discussion can help develop the hypotheses around which a subsequent quasi-experimental input-impact evaluation can be structured. Process research can also provide guidelines for improving inadequate data bases. For example, as a part of the descriptive peer review analysis, the evaluators are in the process of developing a primitive management information system. If implemented, the data gathered through such a system could eventually become the basis for an experimental evaluation of the program itself, or at least subsequent modifications of the program.

This discussion suggests the importance of considering the special methodological requirements which may be placed on research if its purpose is evaluation rather than theory development. Experimental and quasi-experimental methodology is relevant not only to evaluative research, but to research generally. Coleman (1971) has begun to separate out the special requirements of evaluation research, noting the need for decision-oriented, rather than conclusion-oriented, research. By the former, Coleman means research aimed at providing an information basis for social action, rather than research designed to further develop theory about an area of activity. Coleman stresses the importance of timely results and research focused on variables subject to policy manipulation.

A methodology of evaluation research must first and foremost be sensitive to the political nature of the environment within which it must be conducted. Among other things, this will affect the method of site selection. Researchers (e.g., Campbell, chapter 6) have commented on the difficulty of making random assignments of individuals or communities to treatments under ameliorative programs. But random assignment aside, particularistic considerations clearly enter into every major program decision. In a recent large demonstration program, it is widely acknowledged that approximately half of the sites were chosen on the merit of their proposals and the other half on political grounds. Even contracts for evaluation research are subject to these considerations. A contract may be granted because it represents an opportunity for committing a professional organization to self-evaluation, even though the organization's technical competence for evaluation research is very low. What type of methodology can insulate evaluation research from such threats?

The political-fiscal cycle also places inordinate strains on the evaluation process. Contracts may be let for a two-year operating program which is to include the gathering of baseline and posttest data. Once the money has been released, it must be spent within the prescribed time period, even if the survey instruments necessary to baseline data collection cannot be cleared. Thus, the baseline data are likely to be collected at the midpoint or even after some of the more significant actions are undertaken. At this point, the posttest data effort will probably be abandoned. Given the frequency of personnel changes and the short-term time frames within which agencies operate, there are institutionalized pressures working against a sustained interrupted time-series evaluation effort. Sometimes conditions are unintentionally created which prevent valid and useful evaluation at the same time that money is appropriated for major evaluation efforts.

Fortunately, not all the special requirements of evaluation research methodology are so difficult. A good example of an intersection of interest between evaluation researchers and policy makers is their mutual preference for panel and pretest-posttest designs. Administrators unskilled in data analysis seem to have little faith in trend analysis and the sample survey. They object to making decisions based on the opinions or information obtained from 1 per cent, 10 per cent, or even 25 per cent of their constituents. The same decision makers objecting to cross-sectional surveys and random sampling feel confident with data obtained from panel and pretest-posttest designs, as well as from a total census approach to data collection. These strong designs appear to decision makers as logical methods for determining the impact of a particular program.

One other potential intersection of interest between policy makers and evaluation researchers is in the use of a "most different systems" research design. There seems to be a pervasive ideology among policy makers that experimentation and local initiative represent the "American Way." It is thought that local institutions will and probably should resist efforts by a federal agency to specify the administration of experimental programs. According to this philosophy, policy makers prefer to select very dissimilar sites in which to test a program; some may include whole states, while others may cover only a county or a city. This most different systems design can be appropriate for establishing the external validity and generalizability of results. But there may be major questions as to the equivalence of both independent and dependent variables across widely different systems. Under such circumstances a most different systems design may make it more difficult to ascertain whether the observed impact of an experimental program is spurious or not. Attention should be paid to possibilities for developing control groups within the context of a most different systems design, and for promoting the use of randomization. Given the preferences of policy makers, it may behoove the evaluation researcher to reassess and further develop this design.

The importance of adapting evaluation research to the needs and preferences of policy makers cannot be overstressed, since there is a real danger of evaluation saturation. Agencies funding the evaluation research described by Weiss and Rein (1970), as well as those funding much of the evaluation research in the health field, are — or should be — getting skeptical about the payoff from such research. While the importance of developing a valid evaluation research methodology cannot be denied, the methodology must be related to the policy and to the political realities of the situation. Without this concern, the experience of evaluation researchers may be very similar to that of operations researchers: decades of research and applications to health problems with extremely little payoff in terms of actual effects upon operations (Stimson and Stimson 1971).

THREATS DUE TO DIFFICULTIES OF IMPLEMENTING RESEARCH RESULTS

Relatively little attention has been paid to methods for implementing evaluation research results. Yet, from the perspective of the evaluation model, this element is as important as all other elements combined. Fortunately, there are several research traditions which can tell us where to look for answers to this set of problems. Among those that are relevant are diffusion of innovation (Coleman, Katz, and Menzel 1966), integration of operations research solutions into industry (Rubenstein et al. 1967), and, more generally, the organizational change literature (Argyris 1971; Bennis 1969). This literature helps focus on the behavioral and organizational problems of putting into practice conclusions drawn from evaluation research.

Many of the problems observed by evaluation researchers are not unique to any particular discipline. For example, Campbell (chapter 6) has suggested that evaluation research is difficult because administrators advocate programs as though they were certain to be successful, refusing to acknowledge that they could be unsuccessful. Ambiguity, incomparability, and qualitative data are used to give the administrator control over whatever evaluation gets done.

Argyris' (1971) work on the introduction of management information systems into industry suggests that Campbell's observations may apply to the management process generally. Argyris argues that decision makers resist the introduction of a management information system because it emphasizes the use of valid information and technical competence, rather than formal powers, to manage organizations. Such a system removes the need for the executive who can make self-fulfilling prophecies. It does not allow him to escape blame for an incorrect decision because of ambiguity about correct information. The similarity of Campbell's and Argyris' observations suggests that techniques developed by organization theorists should be helpful in solving some of the evaluation researcher's problems. Among other things, Argyris stresses the need to improve the relationship between the evaluation researcher and the administrator and to raise the level of their interpersonal competence. Without an appreciation of the utility of evaluation research, the administrator is likely to continue efforts at sabotage.

While administrative ability to sabotage research is a potential threat, an equally important problem is that of organization design. Frequently there is no simple method for feeding the results of evaluation research into the policy process. Evaluation research is typically commissioned by a special center or work group set up for a more general research purpose. The evaluation researcher will submit a final report to his project officer, and he will be encouraged to publish significant results in academic and professional journals. Especially interesting or relevant topics will be circulated in the form of center publications.

The main consumers of such evaluation efforts are not policy makers, but the academic community and, in some instances, the practitioner. While the academician may be spurred on to do cumulative research, and the practitioner may find the descriptive material useful, this diffusion pattern is quite different from that assumed by the evaluation research model. Considerable attention must be paid to this very basic issue: How can evaluation results be directed to decision makers who can make best use of them?

One approach is the data archive suggested previously. A data archive making past evaluations generally accessible would not only increase the possibilities for analysis of policy alternatives, but also increase the likelihood that policy makers' requests could be quickly serviced. A rapid response time is essential, given the time frameworks within which most policy makers operate.

An archive would also help address the problem of anticipating the policy maker's evaluation needs. Frequently an evaluation is concluded at a time when there is no particular political interest in the program area. Published findings may appear, and a final report may be filed, but it will be very difficult two or three years later to retrieve the original evaluation data when the program area becomes of direct policy interest. Retrieving the original data for further analysis is important because typically the obvious questions will not have been addressed in working papers or publications. Without this archive it is likely that at the height of policy interest new evaluations will be commissioned and the results will again become available only after their relevance for policy has passed.

One operating example of a policy archive is the Community Profile Data Center in the Community Health Service of the Department of Health, Education, and Welfare (HEW). This archive has twenty-two active data files which are continually updated. The files include time-series data on AHA-registered hospitals (facilities, personnel, cost, occupancy rates, etc.), long-term care facilities, physicians (type of practice, age, professional activity, certifications), infant and prenatal mortality rates, population estimates, Medicare enrollees, income estimates, and so forth. Data are coded for retrieval at the lowest geographic level, often as small as the census tract or zip code. Types of requests currently serviced by the data bank include those for standard cost data on physician encounters in neighborhood health centers versus the solo practitioner's office and analyses of the use of ambulatory services by migrant workers. The turn-around time for analyzing the relevant data and providing interpretation is measured in days and weeks rather than in years. Since the director of this data center has recently been placed in charge of all evaluation activities for the Community Health Service, a further test of a data archive for evaluation purposes will be possible.

However, even if evaluation efforts can be designed to feed data into the appropriate policy level, organization research indicates that a learning cycle must be completed (Rubenstein et al. 1967). Often the uses to which evaluation data can be put do not exist until the data exist. In the absence of such data, policy makers rely on other types of indicators or they just do not do what might be possible given the data. In other words, evaluation data may not fill an *existing* need. The reverse is also true; agencies with experience in data analysis may come to view evaluation as an integral part of decision-making. Once policy makers become accustomed to using systematic data, they begin to require data before entering into a major planning effort. They rely on data for pinpointing problems, for developing a shared perspective on these problems, and as a resource to be used while seeking changes in the system.

However, to arrive at this point a substantial initial investment must be made. Unfortunately, so much haphazard and unrelated data collection has gone on that nonbelievers view data collection as a devious means to delay action on seemingly obvious problems. This leads to a vicious circle in which data are not on hand so their usefulness cannot be seen; therefore, no real effort is put into data collection; therefore, they are not on hand.

Another threat to implementation of evaluation research is the basic conflict of interest between policy makers and academicians. This is a serious threat, since much of the ongoing evaluation research is under academic auspices. At a recent health conference, university-based researchers presented to demonstration project personnel instruments which they had developed or pilot-tested. The project personnel were expected to take these instruments and gather data which would be useful to them and helpful for evaluating their efforts. At many points, what should have been a useful interchange of advice and information turned into a very unsatisfactory experience for all involved. The people being asked to collect the data wanted to know the precise purpose and usefulness of each item. The academic personnel refused to be pinned down; they worried about overgeneralization. The academics' response was that the project personnel would have to develop their own analysis according to what they were interested in finding out.

Another conflict between academic and administrative personnel occurred during the reporting of research on alternative methods for obtaining health interview data. Comparative statistics on the following strategies were given: personal interview, telephone interview, and the mailed questionnaire. Administrators were given figures on the percentage of interviews obtained by each method and by combinations of methods, the number of logical errors according to method, the time required to complete each method, and internal comparisons of survey responses according to the method used to obtain information.

While the presentation was very interesting from the perspective of survey research methodology, the people being asked to conduct the survey wanted to know which methods should be used. The researcher's answer was, "Well, it depends. If time and money are not a concern. . . ." This type of response was not satisfactory. Administrators had specific questions: What are the costs? What are the tradeoffs? When should one method be used? When would another method be preferable? While the data reported would have been appropriate for publication in *Public Opinion Quarterly*, they did not provide the answers needed to make administrative decisions—even though administrators might have been able to make the calculations on their own from the data gathered.

In another session, an exchange between academicians who had conducted a national health survey and a staff group responsible for advising policy makers on national health insurance was observed. The staff group was formulating a model for examining the impact of different health insurance policies. The staff was operating within a very tight time framework and was trying to get the best data possible for testing its model. It soon became apparent that the academic group had much of what was needed.

The national survey had sophisticated data on utilization behavior under different types of insurance programs, and these data were being validated. Individual self-reports were being checked out by contacting physicians, hospitals, and insurance companies.

Having established that the necessary data had been collected, the staff group kept asking how soon it could get the calculations it needed. The academics were very reluctant to commit themselves. Some processing had been completed and was available for examination. Several months to a year were needed to clean the data and complete the various validation efforts. The academics noted that their contract provided only limited funds, and an elaborate analytical timetable had already been formulated.

The academics' position was: We want to do the best job possible. The staff's unstated position was: We have to develop our model within a strict time limit—some good data will be extremely valuable now, all perfect data will be useless later. Both groups were reasonable and a useful compromise was made: the staff would prepare their requests within the academics' analytic framework and provide additional resources as necessary. The academics agreed to meet the staff's needs as closely and as quickly as possible.

This exchange underlines the importance of organization design for solving implementation problems. When researchers can communicate readily with policy makers and/or when policy makers can get ready access to appropriate data, the data are likely to be used and the conflict of interest can probably be resolved.

Commissioning consulting organizations rather than academic personnel to do evaluation research is also fraught with difficulties. While some consulting organizations do quality work, there are also built-in problems. If the academic tends to be unresponsive to decision makers' needs, the consulting organization is likely to be overly sensitive. Instances have been observed where a consulting organization asked to evaluate a program provides its client with a whitewash which the evaluator assumes, or has been told, the client expects. In some cases this nonobjective role of the supposed evaluator gets carried to extremes. One notable case involved a consulting group which was asked to evaluate community health planning agencies. Certain members of this group were so committed to grass-roots participation that they spent more time organizing than evaluating.

One final point about the implementation of evaluation research might be made. Because of the evaluation research focus upon

monitoring a policy decision already made, the most likely consumers of evaluation research may be watchdog agencies, such as
the General Accounting Office; the various regulatory agencies,
such as the Federal Trade Commission; and, more generally, the
Congress. Certainly the General Accounting Office is the most
active auditor in government today, and it is increasingly interested in performing impact as well as audit evaluations.

No attempt has been made to cover exhaustively the organizational, political, and behavioral threats to evaluation research.
However, this discussion has stressed that these threats are quite
as capable of invalidating and undermining the evaluation process
as are the other threats previously discussed.

PROPOSED GUIDELINES FOR EVALUATION RESEARCH

This paper has discussed various threats to the usefulness and
validity of evaluation research. With these in mind, it is possible
to separate out two different factors to be considered in designing
an effective evaluation strategy. On the one hand, behavioral,
political, structural, and methodological problems must be addressed. The following are necessary for the effective conduct of
evaluation research:

1. The research must address the policy maker's objectives.
2. The alternative or alternatives examined must be operationally relevant.
3. The methodology must provide "valid enough" results.
4. The policy maker must receive the results in a timely,
   intelligible, nonthreatening fashion.

On the other hand, even if these problems are solved, evaluation
strategy must still be matched to the nature of the particular
program.

In the first part of this paper the similarity between the economists' model of decision-making and the input-impact evaluation
model was mentioned. There is a growing literature which attempts to ascertain those conditions under which the economists'
rational model is appropriate and those under which the model
is inadequate and potentially misleading (Mack 1971; Braybrooke
and Lindblom 1963). It has been suggested that, where there are
explicit goals and an adequate understanding of causality, decisions should be made differently than where such conditions are
lacking. In a well-structured problem, decisions can be made by
setting out objectives, examining how different alternatives will

affect these objectives, and by choosing that alternative which best delivers the desired outcome. In an unstructured, multigoal situation, it is more suitable to focus on a strategy which involves incremental gains and which will move toward a satisfactory solution.

Campbell (1970) and others have recognized that there are conditions under which evaluation research is not appropriate. The above discussion of the rational decision model suggests a framework within which these conditions can be systematically examined. Two factors influencing the type of evaluation which is appropriate are the degree of program goal definition and the degree of obtainable knowledge. Goals, or program objectives, were discussed earlier. Goal definition may vary because people disagree about the goals of a given program; because the goals of a program are not fixed, but evolve over time; or because no one is willing or has given the time to defining them. "Degree of obtainable knowledge" means the degree to which the level of knowledge about the policy area to be evaluated is still in the exploratory research phase—the degree to which researchers are still trying to discover what is related to what. Are enough of the important relationships understood so that the program can be designed to test hypotheses in the form of alternative program strategies (Thompson and Rath 1972)? Can the relevant variables be measured? Can the researchers operationalize important program inputs, outputs, and control groups? Are appropriate data available, or do researchers have substantial funds for data creation? What is the time perspective? Are time-series data available, or can the policy maker wait for multiple time points to be sampled? Adequate time and resources will permit data collection on multiple indicators and the replication of the demonstration in additional research sites.

A program with defined goals and a high degree of obtainable knowledge permits a very different evaluation strategy than a program with undefined goals and a low degree of obtainable knowledge. In figure 9.3 we have drawn on Campbell and Stanley (1963) and others in developing an outline of possible variations along the evaluation research design dimension. The upper half of the figure describes the major design alternatives open to the evaluation researcher, moving from very weak designs on the left to strong ones on the right. The lower half of the figure suggests additional means for improving the level of validity by design and analysis.

Descriptive monitoring would include many of the techniques

FIGURE 9.3.   Possible Variations Along the Evaluation Research
               Design Dimension

Process Design				Experimental Design	
I	II	III	IV	V	VI
Descriptive Monitoring	Auditing	Ex Post Facto with Matched Controls	One Group Pretest-Posttest	Pretest-Posttest Control Group Design	Pretest-Posttest Control Group Design with Random Assignment

Additional Increments of Validity Added Through:
I. Multiple indicators, multiple time periods, multiple replications
II. Advanced data analysis: causal modeling, factor analysis, multiple
    regression with cross-validation, accounting for sampling and
    measurement error by various means

suggested by Weiss and Rein (1970). This is, in essence, the informed common-sense approach to evaluation which is deeply
ingrained in many decision makers. Decision makers are drawn to
this approach because it acknowledges the complexities of the
day-to-day world with which they must deal. They favor or allow
simultaneous multiple changes to be introduced because they
argue that controlled, staged interventions appropriate for captur-
ing all relevant variables would take an eternity. Descriptive
monitoring relies on judgment to synthesize this complexity.
Scientists argue that two equally informed experts will draw
different conclusions from the same chaotic variety, and that it is
therefore necessary to structure the observations in a systematic
way.

Monitoring becomes appropriate where there is little agreement
upon what the objectives of the program are, where there is little
understanding of the effect a program is likely to have, or where
the anticipated effects are so wide-ranging that only an arbitrary
focus would be possible. In all of these cases, the researcher does
not know what data to collect. Monitoring is also applicable where
the type of data wanted is known, but where such data are not
obtainable.

Auditing is another activity which is widely thought of as a useful evaluation effort, even though it does not permit evaluation in
the input-impact sense. Auditing is the activity for which the General Accounting Office is best known. Like monitoring, it focuses on

the process by which a program is conducted. Auditing might focus on the efficient use of resources within a program, the qualifications of a staff for the tasks assumed, or the adherence to program guidelines. Auditing focuses on the efficiency with which a program is being conducted, rather than on the effectiveness with which the program is achieving its objectives.

Ex post facto designs are typically used when input-impact evaluation is wanted but preintervention data are not available. If postintervention data are available on individuals or sites which have not received the treatment as well as on those who have, statistical controls are used in an effort to equate the two groups. There are major problems with the design — namely, the potential for regression artifacts. Campbell and Erlebacher (1970) have forcefully argued that having no evaluation at all is better than relying on this design. This assertion is denied by Evans and Schiller (1970), two administrators closely associated with the ex post facto evaluation of Project Head Start. They grant Campbell and Erlebacher the technical correctness of their argument, while maintaining that the design provided "enough validity" for drawing policy conclusions. This exchange vividly illustrates the tradeoffs between usefulness and validity discussed earlier.

The one-group pretest-posttest design (IV) becomes possible when comparable data on relevant objectives are available both before and after the planned intervention. If such data can be obtained over a number of points in time, this becomes the interrupted time-series design. Design V is possible when pre- and posttest data are available on a comparable group not receiving the intervention, while design VI builds in random assignment.

Clearly, for input-impact evaluation, stronger designs are preferable when possible. While descriptive monitoring, auditing, and ex post facto analysis can supplement the stronger designs, the latter should be used when goals can be identified and time-series data can be compiled. But having said this, one is forced to ask why almost all evaluation relies on the weaker designs. Campbell (chapter 6) suggests that hardheaded evaluation is difficult to implement and that it poses a political threat that makes it unpalatable.

While this is undoubtedly part of the explanation, two other factors seem particularly important. One is a fundamental lack of understanding of evaluation research among top-level policy makers. This is complicated by an inadequate supply of individuals skilled in the fundamentals of research design and appreciative of the dynamics of policy planning. A second factor is the extreme difficulty of anticipating policy makers' evaluation needs. Priorities change quickly in the political arena; it is difficult to predict when and how an issue will develop the critical mass necessary to turn it into a legislative proposal.

POLICY MAKERS' FAILURE TO UNDERSTAND INPUT-IMPACT EVALUATION

There is some indication that policy makers are beginning to appreciate the need for experimentation before making major legislative and monetary commitments. Housing allowance experiments are being tried prior to legislative action in this area, and the recent social security bill, H.R. 1, calls explicitly for experimentation to take place in several policy areas. Unfortunately, well-intentioned administrators frequently assume that evaluation is an activity occurring after a major experimental program has been initiated, rather than an activity which can and should be deliberately designed into an experiment from the beginning. Given the foresight, time, and resources, an understanding of the policy area, and a willingness to specify objectives, experimental sites can be selected according to the main parameters of the program, rather than by default or irrelevant criteria; standardized time-series data can be gathered or sought retrospectively. Given the misunderstanding of evaluation and other pressures contrary to such explicitness, planned experimentation will only come when agencies rethink and restructure their activities. But this is possible, and, with evaluation becoming such a political "buzz" word, agencies may well move or be pushed in this direction.

An alternative role for evaluation would be to design evaluation feedback mechanisms into major legislation, once the basic policy decisions have been made. This is a less dramatic role for evaluation and a much less appreciated one. In the recent social security bill, H.R. 1, innumerable policy objectives have been written into law. Legal mechanisms for obtaining compliance with the law will undoubtedly be developed. However, little or no apparent thought has been given to implementing the bill in such a way that policy makers will be able to tell whether or not the specific objectives of the legislation are being achieved. For example, Section 221 develops a mechanism for forcing hospitals and other providers of health services to have their plans for making major capital expenditures, changing their bed capacity, or changing their service patterns approved by state and local planning agencies. Presumably the objective of the bill is to rationalize the distribution of hospital beds and services. However, there is no indication that the policy makers have appreciated the merits of staged implementation (Campbell, chapter 6) or of a data gathering activity which would permit interrupted time-series analysis of the section's impact. Either of these, but especially the latter, could easily be developed if administrators were aware of the importance and feasibility of such evaluation activity. Good evaluation requires the collection of data over time; because of this requirement and due to the difficulty of foreseeing policy makers' needs, evaluation may be more feasible as a feedback supplement to legislation than as research preceding major action.

ANTICIPATING POLICY MAKERS' REQUIREMENTS

The difficulty of anticipating policy makers' needs becomes a major barrier to the evaluation of experimental programs. This is graphically illustrated by Williams' (1972) discussion of the negative income tax experiments. He notes that "the belief that any serious consideration of the adoption of the negative tax was far in the future was one factor throwing the project well behind schedule so that information was not available at the crisis point of decision-making." He also mentions that the researchers did not correctly anticipate the objectives which would be uppermost in policy makers' minds. The experiment had not addressed itself to how the negative income tax should be integrated with other transfer systems, such as food stamps and rent supplements, or to how a negative tax would affect women.

THE OMNIPRESENCE OF "CATCH-UP" EVALUATIONS

Two basic strategies for evaluation research are evident: the planned strategy which develops time-series data collection and appropriate control groups, and the "catch-up" strategy which relies on the weaker process and ex post facto designs. The "catch-up" strategy occurs because of two factors discussed earlier: policy makers' misunderstanding of evaluation research and the unanticipated nature of policy needs. This suggests that the "catch-up" strategy is going to be very difficult to eliminate totally.

Campbell and Erlebacher (1970) take the position that social scientists should refuse to get involved with weak evaluation efforts, since the results generated by such research are likely to be uninterpretable or misleading. However, the perspective of the middle-level policy maker and evaluator (many of whom have had social science training) must, almost inevitably, differ from this. The policy maker is restricted by the requirements of his role, and is able to concentrate only on improving what is being done in a specific sphere of responsibility. He has very limited control over what sites are selected; once they are selected, he can only encourage and cajole participants into collecting standardized data. These efforts can be easily thwarted. "Dry-labbed" data may be returned, or a call may be placed to his superior or to a congressman concerning federal meddling or invasion of privacy.

The policy maker favoring stronger research designs will probably find it most effective to adopt an incremental strategy: trying to win concurrence on standardized data at one point, time-series data at another, and perhaps controlled site selection at another.

Questions of strategies for organizational change are also involved here. Advocacy by committees of high-level social scientists (Campbell and Erlebacher 1970) may be helpful in bringing about change in evaluation strategies. However, there is a big gap between advocating and doing (Benveniste 1972), and the active collaboration of these social scientists in the design of ongoing evaluation is likely to be much more effective.

It goes without saying that we are a long way from achieving the experimenting society. Hopefully this paper will make some contribution toward the design of more useful and valid evaluation efforts, and thus help to move society in the experimenting direction.

APPENDIX

A ROTATION DESIGN
Administrative Influence
in Turkey

LESLIE L. ROOS, JR.

"Panels, Rotation, and Events" (chapter 2) treated the rotation design, but there have been few examples of this design. Rather than present an extended example of a rotation design in an already long overview essay, I decided to deal with this example more fully here. The logic of the analysis parallels the more general discussion in the overview.

My use of data on Turkish district governors rotated among settings in the 1956 (t_1) to 1962 (t_2) period provides the example, but analysis becomes rather complicated. This research was directed toward ascertaining the effect of setting variables (district characteristics) and individual leader variables (administrator characteristics) upon follower variables (attitudes and orientations of village headmen). The followers' attitudes and behaviors were taken as the important dependent variables, since their position is formally subordinate and their behavior can only gradually feed back to change the settings.

Available survey data (described in Roos and Roos [1971]) were sorted in line with the model presented in figure A.1. In this study, individual administrators were rotated among settings; both setting and individual matches are diagramed in figure A.1. Problems with missing data necessitate presenting the units as in figure A.2. The logic of the setting match can be contrasted with that of the individual match. The time of data collection must be mentioned here; missing data problems did not permit complete application of the relevant methodologies. Essentially, t_1 data from settings and leaders were correlated with t_2 data from followers for both setting and individual matches. The dotted lines in figure A.4 indi-

FIGURE A.1.　Idealized Sorts for Setting and Individual Matches with Followers

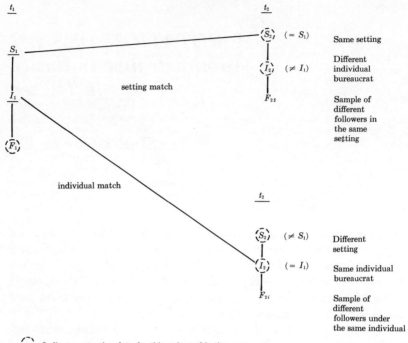

		Same setting
(S_2)	$(= S_1)$	
(I_2)	$(\neq I_1)$	Different individual bureaucrat
F_{2s}		Sample of different followers in the same setting

setting match

individual match

t_2

		Different setting
(S_2)	$(\neq S_1)$	
(I_2)	$(= I_1)$	Same individual bureaucrat
F_{2i}		Sample of different followers under the same individual

$(\ \ )$ Indicates extensive data for this unit at this time were not available.

FIGURE A.2.　Setting and Individual Matches in a Rotation Design

Setting Match	Individual Match
S	S
I	I
F	F

Individual bureaucrats "randomly" distributed. Settings and followers linked.	Settings "randomly" distributed. Individual bureaucrats and followers linked.

S = Setting
I = Individual Bureaucrat
F = Followers

cate correlations which, if assignment were truly random, would be expected to be quite small.

Several plausible causal patterns are presented in figure A.3. One or more characteristics of a setting might directly affect the individual administrator's behavior and/or the follower's attitudes. The characteristics of the setting may also affect the followers' attitudes indirectly by influencing the administrators' behavior, which in turn can affect the followers. Particularly if bureaucrats were not assigned to settings completely randomly, synchronous correlations among settings, leaders, and followers might be found for any of the models outlined in figure A.3. In the setting match, where individuals are more or less randomly rotated between t_1 and t_2, these causal models imply fairly substantial lagged correlations among setting variables (A) and follower variables (C). There may also be moderate lagged correlations between bureaucrat (B) and follower (C) variables; such correlations are suggested in Model II and may be present in Model I. On the other hand, in the individual match where settings are more or less randomized between t_1 and t_2, Models I and II imply low or nonexistent lagged correlations among setting variables (A), bureaucrat variables (B), and follower variables (C).

Different correlational patterns are predicted by Model III,

FIGURE A.3.   Some Possible Causal Models for Setting, Individual Bureaucrat, and Follower Variables

Model I Direct Effect of Setting Variables	Model II Indirect Effect of Setting Variables	Model III Direct Effect of Individual Bureaucrat Variables

A = Setting Variables

B = Individual Bureaucrat Variables

C = Follower Variables

Setting variables affect follower variables directly. Setting variables may affect bureaucrat variables.

Setting variables affect bureaucrat variables which, in turn, affect follower variables.

Bureaucrat variables are independent of setting.

which assumes that attitudes and behaviors of individual administrators are important direct determinants of follower attitudes. In the setting match, with the bureaucrats rotated between t_1 and t_2, low or nonexistent lagged correlations are expected among the three sets of variables. But with the individual match, substantial lagged correlations between the bureaucrat variables (B) and the follower variables (C) are expected. Thus, the patterns of correlations should help in distinguishing among the different causal models.

Having described the research operations and outlined some possible causal models, three sources of error—nonresponse bias, selective leaving, and selective assignment—might be discussed. Nonresponse bias can make both the setting and the individual data unrepresentative of villagers and bureaucrats as a whole. Sampling strategies adopted for both studies were designed to make them as representative as possible (Roos and Roos 1971). Problems of nonresponse could result from the 1956 (t_1) survey of administrators, where the effective response rate was 62 per cent. Selective leaving might also cause difficulties. Since a number of 1956 (t_1) respondents left their posts before the 1962 (t_2) village survey, the administrator-villager combinations on the individual match may be nonrepresentative of both administrators and villagers.

Some idea of the difficulties caused by nonresponse bias and selective leaving can be had by comparing marginals from three groups: village headmen in the 1962 study as a whole, those in the setting match data, and those in the individual match data. Since possible nonresponse was slight in the 1962 village study (95 per cent response rate) and the selective leaving problem is not relevant to the village data considered alone, the village marginals were used as a check on bias in the individual and setting match data. Various comparisons across twenty-three items revealed very few differences in marginals; thus, the results are likely to have been generalizable to the respective universes (Turkish village headmen and district governors) at the time the studies were done (see the extensive discussion in Roos and Roos [1971]).

The issue of random versus selective assignment must be treated. This question is salient substantively because, although regulations promulgated in 1961 called for an equitable rotation of district governors, the assignment process was a constant source of comment and complaint within the Turkish bureaucracy.

Randomness in assignment was first checked by comparing the bureaucrats' 1956 (P_1) and 1962 (P_2) assignments; there was no direct relationship between P_1 and P_2. Next, several variables from the (1956) bureaucrat data were used in an effort to predict 1962 assignment. These variables were chosen from the beta weights assigned to each of a number of variables when the individual match data were analyzed. As McNemar (1969:207) notes, such selection of important variables from a pool "tends to capitalize on correlations which might be high because of sampling fluctuations." McNemar suggests cross-validation procedures to check on the multiple regression equation generated from the initial sample. Specifically, a second sample might be used to calculate a new set of regression weights and a multiple correlation coefficient for the selected predictor variables. Although there is some overlap between the matches, the setting match data provide a second sample which can be used for these cross-validation procedures.

This cross-validation procedure produced some interesting results when the same five predictor variables were run against 1956 assignment (ranking of district where administrator was stationed). For the individual match, the multiple correlation was 0.51, with a highly significant F of 5.91 for 91 cases. For the setting match (the cross-validation sample), the multiple correlation was only 0.18, with a nonsignificant F of 1.47 for 218 cases.

Additional tests using regional rather than district data are available for 1956, 1962, and 1963 assignments. Relationships here were weaker than those noted for the district data; multiple correlations for the cross-validation sample were not statistically significant. Examination of individual bivariate correlations for 1956 and 1962 suggest a statistically significant, but weak, correlation between the administrator's place of birth and his assignment. Bureaucrats from more advanced regions were disproportionately likely to be working in such areas.

Such findings are particularly interesting because, as McNemar (1969:209) notes, the multiple correlation coefficient from the second sample (the setting match in this case) provides "a far better basis for saying what proportion of the variance is attributable to the (selected) independent variables. . . . At this second stage there is no selection that can capitalize on chance." Thus it would appear that—at least insofar as it can be ascertained from measured predictor variables—the bias in assignment was relatively small.

In this analysis, canonical correlations were used to correlate the different sets of measurements in each category (A [setting], B [individual administrator], and C [follower]). This technique has been summarized by Cooley and Lohnes as follows:

The canonical correlation is the maximum correlation between linear functions of the two vector variables. However, after that pair of linear functions that maximally correlates has been located, there may be an opportunity to locate additional pairs of functions that maximally correlate, subject to the restriction that the functions in each new pair must be uncorrelated with all previously located functions in both domains. (1971:169)

Although the empiricism inherent in the use of canonical correlations poses some problems for the development of theory (Blalock 1969:42), this technique seems particularly appropriate for summarizing the relationships among different categories of variables.

As is seen in figure A.4, the data do not permit easy interpretation. Several reasons for these correlations might be operative.

FIGURE A.4.   Canonical Correlations for Setting and Individual Matches

Setting Match	Setting Match	Individual Match
$N = 218$ followers (under 141 bureaucrats)	$N = 91$ followers (randomly selected from the total of 218)	$N = 91$ followers (under 61 bureaucrats)

              A A A

.65, .50, .46   .87, .70, .64   .85, .81, .73

       B B B

         .72, .65, .62   .94, .89, .88   .94, .92, .90

         C C C

A = setting variables
B = individual bureaucrat variables
C = follower variables

The first three canonical correlations are presented next to the line designating the pair of data sets used. There were 6 variables (A) available from the setting data, 22 variables (B) from the individual bureaucrat data, and 36 variables (C) from the follower data. The number of canonical variables generated is equal to the number of variables in the smaller rank of the pair of data sets. All correlations presented here were statistically significant.

Dotted lines indicate correlations which, if assignment were truly random, should be quite small.

The analysis is based on the assumption that setting (A) and bureaucrat (B) variables were similar at t_1 and t_2. Checks on this assumption for bureaucrat variables are provided in Roos and Roos (1971:251).

1. A reason based on the method: canonical correlations were greater for the leader-follower pairs than for the setting-follower pairs because more leader variables than setting variables were used. There were more possibilities for large correlations to be generated by chance.

2. A reason based on the interaction between measure and data: canonical correlations were higher for the individual match because, with a smaller sample (N of 91 as compared with N of 218), sampling and measurement error inflated the correlations obtained by this multivariate procedure. The leader data in particular were know to have been generated with considerable error. Moreover, although means across the two samples were generally quite close, the correlations of individual items with the canonical variables differed substantially between individual and setting matches. As figure A.4 shows, a randomly selected sample of 91 followers from the setting match produced a set of canonical correlations closely resembling the correlations generated by the 91 followers from the individual match. Thus, in this case the over-all differences between individual and setting matches seem to be a function of differing Ns.

3. A reason based on the data: canonical correlations were similar for the leader-follower and setting-follower data because of substantial bias in assignment. Insofar as can be determined from available data, the selection bias was not great; thus this reason seems implausible.

Although these particular findings were not conclusive, the general importance of this line of research should be stressed. In "developed" as well as in developing societies, contact with public officials provides a crucial link between citizen and government. The type of design suggested here is relevant not only for the evaluation of officials but for studying a whole range of questions concerning governmental outputs and outcomes (Jacob 1972).

GLOSSARY

ACTUARIAL APPROACHES employ records collected over time as sources of data. Since the record-keeping process in an institution is often considered a "normal" function of the organization, there is often less chance for the measurement process to be reactive (i.e., the effect of being tested does not change the scores of the subjects with such unobtrusive or nonreactive indicators). (See Webb et al. [1966].)

ADAPTIVE SYSTEMS (see SYSTEMS)

AGGREGATE ANALYSIS deals with data on collectivities of actors; it can be viewed as the "macro" approach to social analysis as opposed to a more individual level of analysis or "micro" orientation. See Gillespie and Nesvold (1971) for examples; see Dogan and Rokkan (1969) for a discussion of the relative benefits of each type of approach.

ATTENUATION OF CORRELATIONS occurs due to the effects of measurement error. The impact of error is to reduce the observed bivariate correlations from what would be obtained with perfect measurement. Various statistical approaches exist (Nunnally 1967:203-4) to estimate the correlations if only true scores were involved. For a discussion of when to correct for attenuation and the effects of correcting with partial correlations, see Bohrnstedt (1969:123-27). See also TRUE SCORE.

AUTOCORRELATION refers to the tendency for temporally successive values of variables in time series to be correlated; this correlation presents problems of a statistical nature since it violates the "independence of observations" assumptions of traditional least squares regression and correlation (Johnston 1972:243-65).

AUTOREGRESSIVE approaches involve regressing values of time-series data upon previous values of the same time series. The logic involved is that certain factors which affect the course of the time series (e.g., certain types of disturbances) may affect the series for more than one observation. Refer to Fox (1968:155-56) for further discussion.

BEHAVIORALISM is a movement in political science which emphasizes the scientific study of politics in terms of behavior. Often this
is accompanied by a stress on quantitative approaches to measurement. The term is used in contrast to more traditional historical
and legal institution-oriented methodologies which once dominated the field. The term should not be confused with BEHAVIORISM
—a totally distinct movement in psychology. See Knorr and Rosenau (1969) for a discussion of the two movements in international
politics; Dahl (1961) is a classic statement dealing with the approach.

BETA COEFFICIENT (see LEAST SQUARES)

BLUESs are the Best Linear Unbiased Estimators of Scalar covariance matrices. A highly technical statistical technique,
BLUESs are employed in the analysis of autocorrelation (Johnston 1972: 254–65). Refer also to AUTOCORRELATION.

BOUNDARIES (see SYSTEMS)

CANONICAL CORRELATION is a statistical technique which relates
more than one predictor variable to more than one criterion. Cooley
and Lohnes (1971:ch. 6) give a detailed introduction.

CAUSAL MODELING (also known as Simon-Blalock causal modeling)
deals with the question of causality by focusing on partial regression and correlation coefficients dealing with (in most cases) synchronous measurements. A major literature has grown up dealing
with this area; see Blalock (1971) for a good over-all introduction;
see also Hilton (1972) and Brewer, Campbell, and Crano (1970)
for criticisms dealing with the topic.

CHANCE REGION (see "TRUE CHANGE")

CHANGE OF STATE (see STATIC)

CLOSED SYSTEM (see SYSTEMS)

COMPLEMENTARY OPERATIONALIZATION is a methodological approach which contends that each method of measuring a phenomenon has unique theoretical implications, hence methods are not as
theoretically neutral as is assumed in convergent (multiple) oper-

ationalism. See chapter 1 of this volume and OPERATIONALIZA-
TION.

CONSTRUCT VALIDITY (see VALIDATION)

CONTENT VALIDITY (see VALIDATION)

CONTINUOUS VARIABLES (see DISCRETE VARIABLES)

CONVERGENT VALIDATION (see VALIDATION)

CORRELATED INDEPENDENT VARIABLES (see MULTICOLLINEARITY)

CORRELATION (see LEAST SQUARES)

CRITERION VALIDITY (see VALIDATION)

CRITERION VARIABLE (see INDEPENDENT VARIABLES)

CROSS-LAGGED PANEL CORRELATIONS (see PANEL STUDIES)

CROSS-SECTIONAL (see LONGITUDINAL)

CYBERNETICS (see SYSTEMS)

CYCLE (see TIME SERIES)

CYCLICAL (see TREND)

DEDUCTIVE (see INDUCTIVE)

DEPENDENT VARIABLE (see LONGITUDINAL)

DETREND means to remove the effects of trend from a time series;
see TREND and TIME SERIES.

DIACHRONIC (see INDEPENDENT VARIABLES)

DIRECTION OF CAUSALITY (see PANEL STUDIES)

DISCRETE VARIABLES in quasi-experimental analysis are the
"events" or singular occurrences which influence subsequent time-

series values. They can be contrasted with continuous variables which are constantly present, in varying value or degree, and which can be used for the paired observations needed for correlational analysis.

DISCRIMINANT VALIDITY (see VALIDATION)

DUMMY VARIABLES (see MEASUREMENT)

DYNAMIC (see STATIC)

ENVIRONMENT (see SYSTEMS)

ERROR means many things to social scientists. Sampling error refers to the fact that repeated random samples from the same population will result in slight differences between the scores obtained on successive, independent samples; tests of significance exist to deal with this problem (see also LEAST SQUARES). Measurement error deals with the notion that all indicators are imperfect (see TRUE SCORE). Variable error results from excluding salient variables from one's equations.

EVENTS (see DISCRETE VARIABLES)

EXPERIMENTAL MORTALITY refers to the problem of subjects who drop out of (or are unlocatable in) a longitudinal study. If these mortality rates are differential (affecting some groups or categories more than others) they can bias outcomes.

EXPERIMENTS are commonly denoted as observations which are subject to manipulation by the researcher and where controls exist for extraneous causal factors. See chapter 1 of this book for an extensive discussion of the meaning of experiment.

EX POST FACTO EXPERIMENTS are attempts to simulate true experiments by matching subjects on pretest scores. Such designs are seriously invalidated by regression artifacts which vitiate the analysis. See Campbell and Stanley (1966) and Campbell and Erlebacher (1970) for further discussion of ex post facto designs and the fallacy of nonrandom matching.

EXTERNAL VALIDITY (see VALIDATION)

FEEDBACK (see SYSTEMS)

FIELD EXPERIMENTS involve the manipulation of at least one independent variable by an experimenter in a natural social setting. They differ from true experiments in that the controls (given by the artificial laboratory setting and the ability to randomize) found in true experiments are not present, hence they are subject to invalidation due to the effects of context. They differ from so-called natural experiments in that manipulation by the researcher takes place. See chapter 1 for a more extensive treatment.

FORM OF A RELATIONSHIP (see LEAST SQUARES)

GAME THEORY (or the theory of games) is an approach to modeling social behaviors originated by Von Neumann and Morgenstern (1947). Defining characteristics of a game (Rapoport 1964:34–37) include two or more players with at least partially conflicting interests; players possessing ranges of choices (termed strategies); simultaneous moves by players; a set of payoffs associated with each outcome of player moves; and having the determination of the game wholly dependent on the player choices. Games are usually less elaborate and more abstract (in terms of their correspondence with the referent universe) than simulations. See SIMULATIONS.

GENERAL SYSTEMS THEORY is a theoretical system that aims at unifying all of science around the concept of systems (see SYSTEMS). The journal *General Systems* indicates the style of work going on in this area. One focal point of the movement has been the search for isomorphisms (or homomorphisms), similarities in structures which exist across content and disciplinary domains.

INDEPENDENT VARIABLES are those whose values are manipulated or whose changes are viewed as the causal factors involved in an experimental or quasi-experimental situation. The dependent variable is the one on whose values the effects of manipulating the independent variable are assessed. In nonmanipulative situations the terms *predictor* and *criterion variable* are often used interchangeably with the independent and dependent variable, respectively.

INDICATORS are often differentiated from variables. The former are the imperfect measurement devices used to make assessments concerning the values of the variable, which is an underlying construct or true score.

INDUCTIVE reasoning proceeds from particulars (observations) to generalizations; in so doing it commits the fallacy of induction since inferences cannot (in any absolute sense) be proven inductively (Campbell and Stanley 1966:17). In evolutionary or trial and error approaches to knowledge (e.g., Campbell 1959; Popper 1959, 1963) this problem is partially avoided by viewing the process of knowing as one involving iterative reality testing: one formulates expectations, tests them in a referent universe, corrects where necessary and reformulates, tests again, etc. Deductive logic goes from general rules to particulars; unlike inductive logic, the findings generated in this manner can be "proven" in that they are completely determined by the rules of logic and the axioms employed. Euclidean geometry is the classic deductive system.

INTERACTIVE processes involve situations (systems) where two or more elements (or actors) are mutually interrelated; the actions or values of one affect the actions/values of the other. This situation is characteristic of systems (see SYSTEMS). This use of the term should not be confused with STATISTICAL INTERACTION, which is a more technical usage.

INTERCEPT (see LEAST SQUARES)

INTERNAL VALIDITY (see VALIDATION)

LAGGED CORRELATIONS involve relating variables measured at different points in time. The units of time difference are the lags (e.g., one year). This technique is often employed in econometric analyses where delayed effects are theoretically expected (e.g., last year's investment affecting this year's growth). Refer to Johnston (1972:292–320) for a technical discussion of some of the statistical problems involved.

LEAST SQUARES is a statistical criterion used in fitting a line to a set of data points. Given two variables (X and Y), we can relate them in a number of systematic ways. Assuming interval-level measurement (see MEASUREMENT) we can produce a scatter plot of the values of the two. Let us assume X is the independent or predictor variable.

We can see that there seems to be a pattern to the cross-plotted dots. By the method of least squares we can fit a straight line to the data with the property that the sum of the squared deviations of

the points from this line will be minimized for the *Y* values; i.e., there is no better fitting line (Blalock 1960:281). This line can be used to predict values of *Y* from values of *X*. The formula is: $Y = a + bX$ where *a* is the intercept (or level, the point where the line crosses the *Y* axis) and *b* is the slope (the rate of change or angle of increase) of the line. Since not all points lie on the line, there will be some error in predictions; the difference between a prediction and the actual score is termed a residual. The entire formula is called the regression equation or form of the relationship. The slope of the line indicates how many units of increase in *Y* are associated with a unit increase in *X*.

In addition to looking at the form of the bivariate relationship, we can look at its strength. This quantity is measured by calculating the spread of the points around the line. The resulting statistic is termed the Pearson product-moment correlation coefficient (or *r*), which varies from +1 to −1; a positive value indicates direct association (increases in one associated with increases in the other) and a negative value an inverse relationship (increases in one associated with decreases in the other). Squaring r (r^2) indicates how much variation the two have in common, in percentage terms.

We can also calculate the statistical significance of the slope, regression equation, and correlation. Statistical significance has two meanings. When dealing with random samples from populations, it deals with sampling error or the probability of getting the obtained result by chance. When dealing with entire populations, it helps control for the plausible rival hypothesis that the obtained results could have occurred through a chance pattern of observation. In neither case does statistical significance equal substantive or theoretical importance. See Blalock (1960) or any other standard statistical text for an introduction to least squares regression and correlation.

LEVEL (see STATIC)

LONGITUDINAL or diachronic analysis deals with the values of one entity or variable over two or more time points. This can be contrasted to cross-sectional or synchronous analysis, which deals with the values of two or more entities/variables at the same point in time. Often, given the incremental nature of many social variables over time, this results in the longitudinally analyzed variable having less variation than would result in a cross-sectional study (for example, there is less variation in U.S. GNP over a two-year span than between the U.S. and Chad in any given year). Each has its own values and limits; both can be profitably combined (e.g., in panel studies).

MATCHING is the attempt to nonrandomly equate groups (e.g., a control and experimental treatment set of subjects), often by the use of pretest scores. Such attempts usually fail when applied to individuals due to regression artifacts. See EX POST FACTO EXPERIMENTS and REGRESSION ARTIFACTS.

MEASUREMENT refers to the assigning of number or magnitude to observations according to rules. There are four commonly employed levels of measurement: nominal, where the numbers serve only as labels (1 if male, 0 if female); ordinal, where one quantity can be said to be greater (lesser) than another but where the magnitude of the difference is unspecified; interval, where the magnitude of difference is specified; and ratio, where a natural zero point exists. Each higher level incorporates all of the properties of the lower.

Special note should be made of the notion of dummy variables. There are dichotomous nominal-level variables (e.g., male, non-male) which can be treated as interval scales with only two values.

MEASUREMENT ERROR (see ERROR)

MONOTONIC FUNCTIONS of the bivariate form $Y = fX$ are said to exist if an increase in the value of X is always associated with an increase in the value of Y. Some sources (e.g., Hays 1963) use the term "monotone-increasing" to refer to this case and the label "monotone-decreasing" to distinguish an inverse relationship, where increases in X are always associated with decreases in Y.

MOST DIFFERENT SYSTEMS DESIGNS (see MOST SIMILAR SYSTEMS DESIGNS)

MOST SIMILAR SYSTEMS DESIGNS are presented by Przeworski and Teune (1970:32–34) as the predominant designs in social science dealing with comparative research. Studies employing this design try to match systems on the basis of similarities (e.g., all Scandinavian states). The approach is on a systems level with elements common to all units being viewed as controlled while intersystem differences are regarded as potential explanatory variables. One problem with this design is that, even with the matching, so many intersystemic differences exist as plausible causal factors. The most different systems design takes as its initial level of analysis one lower than system. System-level differences are introduced as causal factors only when differences in subpopulations cannot be explained otherwise; see especially Przeworski and Teune (1970:ch. 2) for further discussion. The most different systems approach assumes (anticipates finding) greater homogeneity of subjects across systems than the most similar systems design.

MULTICOLLINEARITY refers to a statistical problem area in multiple and canonical regression where two or more of the predictor variables are intercorrelated (i.e., nonorthogonal). In such circumstances, the relative importance of the various correlated predictors is difficult to ascertain due to redundancy in the predictor set (Vincent 1971:40–43). Additionally, Johnston (1972:159–68) lists estimation, sampling and error problems. See also Fox (1968:sec. 7.5).

MULTIMETHODS are used in a methodological orientation which holds that any single indicator measures imperfectly and that in order to avoid this limitation multiple methods of measurement (selected so as to maximize the heterogeneity of irrelevancies in each) should be employed to "triangulate in" on variables and concepts. Campbell and Fiske (1959) present both the logic involved and a technique (the multitrait, multimethod matrix) which is central to the approach. Refer also to OPERATIONALIZATION.

MULTIPLE OPERATIONALISM (see MULTIMETHODS and OPERATIONALIZATION)

MULTIPLE REGRESSION is an extension of least squares regression to include two or more predictor variables in relation to a single criterion. See also LEAST SQUARES, MULTICOLLINEARITY.

NONCHANCE CHANGE (see "TRUE CHANGE")

NONREACTIVE MEASURES (see ACTUARIAL APPROACHES)

NONRESPONSE BIAS refers to subjects in a research setting which requires voluntary cooperation such as refusing to return a mailed questionnaire. If some refuse to cooperate, and, as is likely in many settings, this nonresponse rate is biased (affects some groupings more than others), the effect is to underrepresent the low-responding group, with attendant consequences for external validity.

OPEN SYSTEM (see SYSTEMS)

OPERATIONALIZATION is the use of objective, replicable measuring devices to express a variable's value in terms of values of an indicator. Multiple operationalism is the position that each measuring device assesses imperfectly the concept being measured, hence MULTIMETHODS should be employed. This multiple measurement approach can be contrasted with a strict definitional operationalism which holds that any one measurement scheme denotes a concept or variables accurately (Campbell and Fiske 1959). See also VALIDATION.

PANEL STUDIES deal with observations on the same set of subjects at two or more points in time. Panel studies allow the use of cross-lagged panel correlations (Rozelle and Campbell 1969) and other approaches to assessing the preponderant direction of causality (Darlington 1968). The essential idea behind the cross-lagged panel is that a cause at time 1 should correlate more highly with an effect at time 2 than should an effect at time 2 with a cause at time 1.

PARTIAL CORRELATION is a statistical technique which attempts to assess the amount of variation a given independent variable has in common with a dependent variable while "controlling for" the variation of other independent variables which are in common with the dependent: $rAB \cdot C$ is the partial correlation of A and B, partialing out C. Blalock (1960) and other standard texts go into the method in detail. See also Brewer, Campbell, and Crano (1970) for some cautionary observations on the use of this technique.

PATH ANALYSIS (see CAUSAL MODELING)

PAYOFF MATRIX refers to a set of outcomes associated with the results of a game. See GAME THEORY.

PERIODICITY (see TIME SERIES)

PLAUSIBLE RIVAL HYPOTHESES: an orientation toward research espoused by Campbell in chapter 6 of this volume. Plausible rival hypotheses are used in two senses. First, they are those rival explanations that could reasonably be used to deny the results obtained in a particular quasi-experiment. In this sense they can be used to argue *against* the internal and external validity of specific quasi-experiments. The second use is in *favor* of particular quasi-experiments and is directed at perfectionist critics: the only threats to validity admitted are those which would have the status of laws which were more dependable and plausible than the law used and tested in the quasi-experiment. Testing the hypotheses involved in quasi-experimentation is a continuing process of successively rejecting plausible rival hypotheses.

POST-BEHAVIORALISM is the name of a movement in political science which attempts to follow up on BEHAVIORALISM by increasing the policy and social relevance of political research (Easton 1969).

PREDICTOR VARIABLE (see INDEPENDENT VARIABLES)

PREPONDERANT CAUSATION (see PANEL STUDIES)

PRINCIPAL COMPONENTS ANALYSIS is a form of factor analysis (Rummel 1971a:112–13) which determines a number of factors or dimensions which summarize a larger number of variables. The factors are a more parsimonious representation of the patterns of interrelationship existing in the variables. By selecting the first few factors from a principal components analysis which usually account for the vast preponderance of the variance, one can switch from fifty variables to five or six factors, a clear gain in analytic and conceptual ease.

PROCESS (see STATIC)

RANDOM FLUCTUATIONS occur in time series. One usually observes fluctuations around trend lines with such data; these small variations can be due to sampling error, measurement error, or the essentially random effects of other variables. These fluctuations constitute noise for purposes of analysis. To be considered "significant," a quasi-experimentally induced change in the series must be of a larger magnitude than these background effects.

REGRESSION ARTIFACTS are a recurring problem in statistical analyses. If subjects are selected for treatment (e.g., a remedial program) on the basis of their extreme scores on a pretest, then they will move closer (on the average) to the population mean on a posttest, even if the treatment had no effect. This effect is due to the nature of imperfect measurement and lack of perfect correlation between pre- and posttest. The amount of regression to the mean to be expected is the inverse of the correlation of pre- and posttest. A related analytical problem occurs when one is dealing with a non-random control group. If the control group is more or less able (on the average) than the experimental treatment group, then each will regress to its own population mean on the posttest making the treatment look artificially effective or impotent. See chapter 1 of this volume, Campbell and Stanley (1966), and Campbell and Erlebacher (1970) for further discussions of this most serious problem.

RELIABILITY is the agreement of two or more efforts to measure the same thing through maximally similar methods (Campbell and Fiske 1959). Reliability can be assessed over time (e.g., via test-retest correlations) or across subjects synchronously (e.g., through split-halves correlations). A good synonym for reliability is consistency: do parallel (or the same) measuring devices at different points in time and space give the same readings? Bohrnstedt (1970) gives a good introduction. See also VALIDATION.

RESIDUAL (see TREND, LEAST SQUARES)

SAMPLING ERROR (see ERROR)

SEASONAL COMPONENT (see TREND)

SIGNIFICANCE (see LEAST SQUARES)

SIMON-BLALOCK CAUSAL MODELING (see CAUSAL MODELING)

SIMULATIONS consist of models constructed to be homomorphic (or, more rarely, isomorphic) to some referent universe. Some writers (e.g., Raser 1969) distinguish simulations from games by arguing that the latter are more tenuously constructed and have a higher role for human participants.

SLOPES (see TIME SERIES)

SOCIAL STRUCTURES are persistent patterns of interaction in social contexts.

SPURIOUS CORRELATION describes a bivariate correlation which is judged, on the basis of causal modeling (Simon 1954), not to imply a causal connection. If the correlation between A and B virtually disappears when C is controlled for (partialed out), then one can make an argument that A and B are both caused by C and that there is not a causal connection between them. Certain rather specific assumptions have to be made to employ this approach; for instance, one must assume that C is not an intervening variable through which all of the A to B causation must proceed. See Brewer, Campbell, and Crano (1970) for a cautionary discussion of the use of partial correlations.

STANDARDIZED SCORES are found by the formula: standard score = $\frac{\text{raw score} - \text{mean}}{\text{standard deviation}}$. By this formula, the extremity (in relative terms) of any two scores, measured on any metric system, can be directly compared. Standardized scores are also known as z scores.

STATE (see STATIC)

STATIC and dynamic variables can be employed in hypotheses. A static (or level) variable refers to the level of a trait; a dynamic (or process) variable refers to a shift or change in level. Examples would be having an attitude (state) *vs.* having a change in attitude (dynamic) as predictors and having a behavior (static) or a change in behavior (dynamic) as criterion variables.

STATISTICAL INTERACTION is a problem which confronts statistical analysts. It occurs whenever the combination of two values of two variables produces results which would not be predicted on the basis of the values of either, taken separately. This conjoint interdependence is evaluated in terms of its effects on the slopes of the variables and can be a problem for additive regression models. Blalock (1969) provides an introduction to the problem area.

STATISTICAL REGRESSION (see REGRESSION ARTIFACTS)

STATUS FIELD THEORY, also known as social field theory, is a major theoretical position in international relations. Its essential concern is with the distances between entities on attribute dimensions and the effects of these distances on their dyadic interactions. Rummel (1971b) and Park (1972) provide good introductions; Park in particular shows the more general applicability of the theory outside of international relations.

STRENGTH OF A RELATIONSHIP (see LEAST SQUARES)

SUBSYSTEMS (see SYSTEMS)

SYNCHRONOUS (see LONGITUDINAL)

SYSTEMS consist of two or more units or elements which interact. As can be seen, this definition is a rather general one, hence the researcher must specify (in any content domain) the particular range of elements and interactions he wishes to concentrate on (Ashby 1964:40). Systems can be distinguished from the broader social/physical context in which they exist; this context is called the environment. Systems can be affected by the environment's inputs (open systems) or be immune from its influences (closed systems). Open systems are more characteristic of social behaviors. The defined line between a system and its environment is called its boundary; usually this line delimits differential patterns of interaction. Within a given system there can be any number of subsystems; these can be viewed as systems for which the original system serves as part of the environment.

   Cybernetics, "the science of control and communication" (Ashby 1964:1), is an important part of systems thinking. Systems can receive feedback or information from their environments; this can be positive (destabilizing) and negative (stabilizing). A system with balanced positive and negative feedback is in equilibrium. In interacting with their environments, systems can be adaptive, adjusting themselves continuously in a search for equilibrium.

T TEST is one common variety of significance test.

TEST OF SIGNIFICANCE (see LEAST SQUARES)

TIME SERIES refers to a set of observations on an entity over time. Following is an unrealistic heuristic example.

   Time series can exhibit regular patterns, called periodicities. One form of periodicity is trend — an increase or decrease in mean value over time. Our example has a positive trend from 1960 to

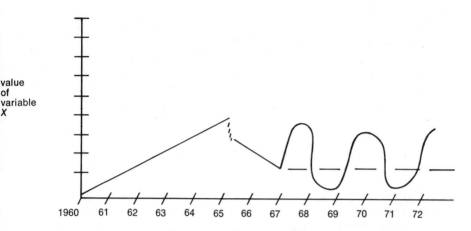

1965. The values of a time series can be affected by disturbances (which may be quasi-experimental treatments). Such an effect may affect either the slope or the mean value (magnitude or intercept) of the series; in our case both changes occurred and persisted from 1965 to 1967. Such series can also exhibit regular patterns of oscillations, one form of which is a cycle (1967+) where peaks (high points) and troughs (lows) are fairly evenly distributed around the mean tendency or trend in the series (broken line).

TREND refers to one component into which econometricians often divide time-series data (the other components being seasonal, cyclical, and residual). Trend (sometimes called secular trend in this context) deals with a general increase or decrease in the series' values over time. Seasonal variation is reflected in patterned fluctuations corresponding to calendar intervals (for instance, upward sales before Christmas). Cyclical variation consists of peaks and troughs, usually of regular pattern. Residual variation is what remains after the other three components have been removed. In dealing with the components of time-series data, it is important to have theoretical reasons for removing components; for example, businessmen often wish to evaluate sales for the same month in different years; here, separation of the trend and cyclical components might be illuminating. See Mason (1967) for a discussion of the components of time series.

"TRUE CHANGE" is one whose shift in value is large enough to exceed the normal fluctuations in time-series data. See also RANDOM FLUCTUATIONS.

TRUE SCORE is a concept utilized in the literature on measurement theory. If we assume that all measurement is imperfect then we can conceptualize a subject's score (X_i) as consisting of measurement of the actual variable (true score) and random measurement error: $X_i = t_i + e_i$. If the randomness assumption is correct, then over the long run (given a few technical assumptions) the expected value of a person's score is his true score (Bohrnstedt 1970). This assumption is not warranted if the subjects have been selected on the basis of their extreme scores; in this case REGRESSION ARTIFACTS will result.

VALIDATION is typically used to refer to the degree to which an instrument (measurement process) measures the concept under analysis: Does it measure what it is designed to measure? There are common subcategories of validity. Convergent validity (Campbell and Fiske 1959) is measured by the relative agreement of independent attempts to measure the same concept which involve maximally dissimilar indicators. The discriminant validity of an instrument is established by showing that a test designed to measure construct A does not highly correlate with other tests designed to measure independent concepts.

A second use of the term refers to content, criterion, and construct validity. Content validity deals with the extent to which measuring devices relate to the concept under analysis. Criterion validity is established by relating the measures employed with direct measures of the characteristic under investigation (e.g., correlating an attitude item scale on political preferences with voting behavior). Construct validity is estimated by relating measurement scores to explanatory constructs. Campbell (1960b) distinguishes between criterion and construct validity by arguing that the former has no a priori defining criterion; instead, independent (convergent) means of getting at the same construct are employed.

A third use of the term contrasts internal and external validity. Internal validity deals with whether a demonstrable change or effect did take place; as such it is fundamental to the evaluation of quasi-experiments. External validity deals with the question of generalizability of findings which have already been shown to be internally valid to other subjects and occasions. See Campbell and Stanley (1966).

VARIABLE ERROR (see ERROR)

VARIABLES (see INDICATORS)

BIBLIOGRAPHY

ACKOFF, RUSSELL L.
    1962   Scientific Method: Optimizing Applied Research Decisions. New York: Wiley.

AGENCE EUROPE
    1964   Agricultural Marathon Well on Way to Compromise Based on Commission's Proposals. No. 2008 (Dec. 24):1.

ALMOND, GABRIEL A., and SIDNEY VERBA
    1963   The Civic Culture. Princeton, N.J.: Princeton University Press.

ALTHAUSER, R. P., T. A. HEBERLEIN, and R. A. SCOTT
    1971   A Causal Assessment of Validity: The Augmented Multitrait-Multimethod Matrix. In Causal Models in the Social Sciences, H. M. Blalock, Jr., ed. Chicago: Aldine-Atherton.

ARGYRIS, C.
    1970   Intervention Theory and Method: A Behavioral Science View. Reading, Mass.: Addison-Wesley.

    1971   Management Information Systems. Management Science 17:B275–B292.

ARMOR, DAVID J.
    1972   The Evidence on Busing. The Public Interest 28:90–126.

    1973   The Double Double Standard: A Reply. The Public Interest 30:119–31.

ARONSON, ELLIOT, and J. MERRILL CARLSMITH
    1968   Experimentation in Social Psychology. In The Handbook of Social Psychology, Gardner Lindzey and Elliot Aronson, eds. 2d ed., vol. 2 (Research Methods). Reading, Mass.: Addison-Wesley.

ASHBY, W. R.
    1964   Introduction to Cybernetics. London: Metheun.

AUBERT, V.
    1959   Chance in Social Affairs. Inquiry 2:1–24.

BALTZELL, E. DIGBY
    1958   Philadelphia Gentlemen: The Making of a National Upper Class. New York: Free Press.

BANKS, ARTHUR S., and PHILLIP M. GREGG
    1965   Grouping Political Systems: Q Factor Analysis of a Cross-Polity Survey. American Behavioral Scientist 9:3–6.

BAUER, R. M.
    1966   Social Indicators. Cambridge: M.I.T. Press.

BAVCHUCK, NICHOLAS, and ALAN BOOTH
1969   Voluntary Association Membership: A Longitudinal Analysis.
        American Sociological Review 34:31–45.

BENNIS, W. G.
1969   Organization Development: Its Nature, Origins, and Prospects.
        Reading, Mass.: Addison-Wesley.

BENVENISTE, G.
1972   The Politics of Expertise. Berkeley, Calif.: The Glendessary
        Press.

BERELSON, BERNARD R., PAUL F. LAZARSFELD, and WILLIAM N. MCPHEE
1954   Voting. Chicago: University of Chicago Press.

BETEILLE, ANDRE, ed.
1969   Social Inequality. Introduction. Middlesex, England: Penguin.

BLALOCK, HUBERT M., JR.
1960   Social Statistics. New York: McGraw-Hill.
1964   Causal Inferences in Non-Experimental Research. Chapel Hill,
        N.C.: University of North Carolina Press.
1967   Causal Inferences, Closed Populations, and Measures of Asso-
        ciation. American Political Science Review 61:130–36.
1969   Theory Construction: From Verbal to Mathematical Formula-
        tions. Englewood Cliffs, N.J.: Prentice-Hall.
1970   A Causal Approach to Nonrandom Measurement Errors.
        American Political Science Review 64:1099–1111.

BLALOCK, HUBERT M., JR., ed.
1971   Causal Models in the Social Sciences. Chicago: Aldine-Atherton.

BLAYNEY, J. R., and I. N. HILL
1967   Fluorine and Dental Caries. Journal of the American Dental
        Association. Special Issue. 74:233–302.

BLISS, HOWARD, ed.
1970   The Political Development of the European Community: A
        Documentary Collection. Waltham, Mass.: Blaisdell.

BLOCK, J., and J. BLOCK
1952   An Interpersonal Experiment on Reactions to Authority. Hu-
        man Relations 5:91–98.

BLUMENTHAL, RICHARD
1969   The Bureaucracy: Antipoverty and the Community Action
        Program. In American Political Institutions and Public Policy,
        Allan P. Sindler, ed. Boston: Little, Brown.

BOHRNSTEDT, GEORGE W.
1969   Observations on the Measurement of Change. In Sociological
        Methodology 1969, E. F. Borgatta, ed. San Francisco, Calif.:
        Jossey-Bass.
1970   Reliability and Validity Assessment in Attitude Measurement.

In Attitude Measurement, G. F. Summers, ed. Chicago: Rand McNally.

BOHRNSTEDT, GEORGE W., and I. M. CARTER
1971   Robustness in Regression Analysis. *In* Sociological Methodology 1971, H. L. Costner, ed. San Francisco, Calif.: Jossey-Bass.

BORGATTA, E. F., ed.
1969   Sociological Methodology 1969. San Francisco, Calif.: Jossey-Bass.

BORING, EDWIN G.
1954   The Nature and History of Experimental Control. American Journal of Psychology 67:573–89.

BORUCH, R. F.
1971   Maintaining Confidentiality of Data in Educational Research: A Systemic Analysis. American Psychologist 26:412–30.

BOUDON, RAYMOND
1968   A New Look at Correlation Analysis. *In* Methodology in Social Research, Hubert M. Blalock, Jr., and Ann B. Blalock, eds. New York: McGraw-Hill.

BOULDING, KENNETH E.
1953   Toward a General Theory of Growth. Canadian Journal of Economics and Political Science 19:326–40.
1970   A Primer on Social Dynamics. New York: Free Press.

BOX, GEORGE E. P.
1967   Bayesian Approaches to Some Bothersome Problems in Data Analysis. *In* Improving Experimental Design and Statistical Analysis, Julian C. Stanley, ed. Chicago: Rand McNally.

BOX, GEORGE E. P., and G. M. JENKINS
1962   Some Statistical Aspects of Adaptive Optimization and Control. Journal of the Royal Statistical Society 24:297–343.

BOX, GEORGE E. P., and GEORGE C. TIAO
1965   A Change in Level of a Non-Stationary Time Series. Biometrika 52:181–92.

BRAYBROOKE, D., and C. LINDBLOM
1963   A Strategy of Decision. New York: Free Press.

BREWER, MARILYN B., DONALD T. CAMPBELL, and WILLIAM D. CRANO
1970   Testing a Single-Factor Model as an Alternative to the Misuse of Partial Correlations in Hypothesis-Testing Research. Sociometry 33:1–11.

BRICKMAN, HOWARD J.
1969   Conceptual Frameworks for the Study of Integration. Mimeographed. Department of Political Science, Northwestern University.

BROWNE, E. C., and M. N. FRANKLIN
1973   Prerequisites of Government: Aspects of Coalition Payoffs in

European Parliamentary Democracies. American Political Science Review 67:453–69.

BRUNNER, RONALD D., and KLAUS LIEPELT
1970   Data Analysis, Process Analysis and System Change. Institute of Public Policy Studies Discussion Paper No. 27, University of Michigan.

BRUYN, SEVERYN
1966   Human Perspective in Sociology: The Methodology of Participant Observation. Englewood Cliffs, N.J.: Prentice-Hall.

BUCKLEY, WALTER, ed.
1967   Sociology and Modern Systems Theory. Englewood Cliffs, N.J.: Prentice-Hall.

1968   Modern Systems Research for the Behavioral Scientist. Chicago: Aldine.

BULLETIN DER EUROPÄISCHEN WIRTSCHAFTSGEMEINSCHAFT
1961   No. 4 (April). Brussels: Commission Secretariat.

BUTLER, D., and D. STOKES
1969   Political Change in Britain: Forces Shaping Electoral Choice. New York: St. Martin's.

CALDWELL, TED W., with LESLIE L. ROOS, JR.
1971   Voluntary Compliance and Pollution Abatement. In The Politics of Ecosuicide, Leslie L. Roos, Jr., ed. New York: Holt, Rinehart and Winston.

CAMPBELL, ANGUS
1966   Surge and Decline: A Study of Electoral Change. In Elections and the Political Order, Angus Campbell et al., eds. New York: Wiley.

CAMPBELL, ANGUS, PHILIP E. CONVERSE, WARREN MILLER, and DONALD STOKES
1960   The American Voter. New York: Wiley.

CAMPBELL, ANGUS, GERALD GURIN, and WARREN E. MILLER
1954   The Voter Decides. Evanston, Ill.: Row Peterson.

CAMPBELL, DONALD T.
1957   Factors Relevant to the Validity of Experiments in Social Settings. Psychological Bulletin 54:297–312.

1959   Methodological Suggestions from a Comparative Psychology of Knowledge Processes. Inquiry 2:152–67.

1960a   Blind Variation and Selective Retention in Creative Thought as in Other Knowledge Processes. Psychological Review 67:380–400.

1960b   Recommendations for APA Test Standards Regarding Construct, Trait, or Discriminant Validity. American Psychologist 15:546–53.

1963   From Description to Experimentation: Interpreting Trends as

Quasi-Experiments. *In* Problems in Measuring Change, C. W. Harris, ed. Madison, Wis.: University of Wisconsin Press.

1965   Common Fate, Similarity, and Other Indices of the Status of Aggregates of Persons as Social Entities. *In* Human Behavior and International Politics, J. David Singer, ed. Chicago: Rand McNally.

1967   Administrative Experimentation, Institutional Records, and Nonreactive Measures. *In* Improving Experimental Design and Statistical Analysis, J. C. Stanley, ed. Chicago: Rand McNally.

1968   Quasi-Experimental Design. *In* International Encyclopedia of the Social Sciences, 5, D. L. Sills, ed., New York: Macmillan and Free Press.

1969a   Definitional versus Multiple Operationism. Et al 2:14–17.

1969b   Objectivity and the Social Locus of Scientific Knowledge. Presidential Address to the Division of Social and Personality Psychology of the American Psychological Association, Washington, D.C.

1970   Considering the Case against Experimental Evaluations of Social Innovations. Administrative Science Quarterly 15: 110–13.

1971   Temporal Changes in Treatment-Effect Correlations: A Quasi-Experimental Model for Institutional Records and Longitudinal Studies. *In* Proceedings of the 1970 Invitational Conference on Testing Problems, G. V. Glass, ed. Princeton, N.J.: Educational Testing Service.

1972   Methods for the Experimenting Society. Mimeographed. Department of Psychology, Northwestern University.

CAMPBELL, DONALD T., and A. ERLEBACHER
1970   How Regression Artifacts in Quasi-Experimental Evaluations Can Mistakenly Make Compensatory Education Look Harmful. *In* Compensatory Education: A National Debate (Vol. 3 of The Disadvantaged Child), J. Hellmuth, ed. New York: Brunner/Mazel.

CAMPBELL, DONALD T., and D. W. FISKE
1959   Convergent and Discriminant Validation by the Multitrait-Multimethod Matrix. Psychological Bulletin 56:81–105.

CAMPBELL, DONALD T., and H. L. ROSS
1968   The Connecticut Crackdown on Speeding: Time-Series Data in Quasi-Experimental Analysis. Law and Society Review 3:33–53.

CAMPBELL, DONALD T., and JULIAN C. STANLEY
1966   Experimental and Quasi-Experimental Designs for Research. Chicago: Rand McNally. Originally appeared as Experimental and Quasi-Experimental Designs for Research on Teaching. *In* Handbook of Research on Teaching, N. L. Gage, ed. Chicago: Rand McNally, 1963.

CAMPS, MIRIAM
    1966   European Unification in the Sixties. New York: McGraw-Hill.
CAPORASO, JAMES A., and A. PELOWSKI
    1971   Economic and Political Integration in Europe: A Time-Series
            Quasi-Experimental Analysis. American Political Science
            Review 65:418–33.
CARNAP, RUDOLPH
    1956   The Methodological Character of Theoretical Concepts. In
            The Foundations of Science and the Concepts of Psychology and
            Psychoanalysis, Herbert Feigl and Michael Scriven, eds.
            Vol. 1 of the Minnesota Studies in the Philosophy of Science.
            Mineapolis: University of Minnesota Press.
CHADWICK, RICHARD
    1972   Theory Development through Simulation: A Comparison and
            Analysis of Associations among Variables in an International
            System and an International Simulation. International Studies
            Quarterly 16:83–127.
CHAPIN, F. S.
    1947   Experimental Design in Sociological Research. New York:
            Harper.
CLAUSEN, AAGE
    1967   Measurement Identity in the Longitudinal Analysis of Legisla-
            tive Voting. American Political Science Review 61:1020–35.
CLOWARD, RICHARD A., and FRANCIS FOX PIVEN
    1971   Welfare for Whom? In Black Politics, Edward G. Greenberg,
            Neal Milner, and David Olson, eds. New York: Holt, Rinehart
            and Winston.
COLEMAN, JAMES S.
    1966   Equality of Educational Opportunity. Washington, D.C.: Gov-
            ernment Printing Office.
    1971   Problems of Conceptualization and Measurement in Studying
            Policy Impacts. Presented at the Conference on the Impacts of
            Public Policies, St. Thomas, U.S. Virgin Islands.
COLEMAN, JAMES S., E. KATZ, and H. MENZEL
    1966   Medical Innovation: A Diffusion Study. Indianapolis, Ind.:
            Bobbs-Merrill.
COLEMAN, JAMES S., and C. G. ROSBERG
    1964   Political Parties and National Integration in Tropical Africa.
            Berkeley, Calif.: University of California Press.
CONVERSE, P. E.
    1964   The Nature of Belief Systems in Mass Publics. In Ideology and
            Discontent, David Apter, ed. New York: Free Press.
    1970   Attitudes and Non-Attitudes: Continuation of a Dialogue. In
            The Quantitative Analysis of Social Problems, E. R. Tufte, ed.
            Reading, Mass.: Addison-Wesley.

COOK, T. J., and F. P. SCIOLI, JR.
1972   A Research Strategy for Analyzing the Impact of Public Policy.
        Administrative Science Quarterly 17:328–39.

COOLEY, WILLIAM W., and PAUL R. LOHNES
1971   Multivariate Data Analysis. New York: Wiley.

COOMBS, CLYDE H.
1964   A Theory of Data. New York: Wiley.

COSER, LEWIS A.
1956   The Functions of Social Conflict. New York: Free Press.

COUNCIL ON ECONOMIC PRIORITIES
1970   Paper Profits: Pollution in the Pulp and Paper Industry. New
        York: Council on Economic Priorities.

CRAINE, L. F.
1971   Institutions for Managing Lakes and Bays. Natural Resources
        Journal 11:519–46.

CRANO, WILLIAM D., DAVID A. KENNY, and DONALD T. CAMPBELL
1972   Does Intelligence Cause Achievement?: A Cross-Lagged Panel
        Analysis. Journal of Educational Psychology 53:258–75.

CRONBACH, LEE, and LITA FURBY
1970   How Should We Measure Change – Or Should We? Psychologi-
        cal Bulletin 74:68–80.

DAHL, R.
1961   The Behavioral Approach in Political Science: Epitaph for a
        Monument to a Successful Protest. American Political Science
        Review 55:763–72.

DARLINGTON, R. B.
1968   Multiple Regression in Psychological Research and Practice.
        Psychological Bulletin 69:161–82.

DAVID, STEPHAN M.
1968   Leadership of the Poor in Poverty Programs. Academy of Polit-
        ical Science Proceedings 29:86–89.

DAVIS, KINGSLEY, and WILBERT MOORE
1949   Some Principles of Stratification. *In* Sociological Analysis, L.
        Wilson and W. L. Kolb, eds. New York: Harcourt, Brace.

DEAN, DWIGHT
1960   Alienation and Political Apathy. Social Forces 38:185–89.

1961   Alienation: Its Meaning and Measurement. American Socio-
        logical Review 26:753–58.

DENZIN, NORMAN
1971   The Logic of Naturalistic Inquiry. Social Forces 50:166–82.

DE TOQUEVILLE, ALEXIS
1955   The Old Regime and the French Revolution. Trans. Stuart
        Gilbert. New York: Doubleday.

DEUTSCH, KARL
　1961　Social Mobilization and Political Development. American Polit-
　　　　ical Science Review 55:493–514.
DEUTSCH, KARL, SIDNEY BURRELL, et al.
　1966　Political Community and the North Atlantic Area. *In* Interna-
　　　　tional Political Communities: An Anthology. New York:
　　　　Anchor.
DEWEY, JOHN, and ARTHUR F. BENTLEY
　1949　Knowing and the Known. Boston: Beacon.
DEXTER, LEWIS ANTHONY
　1970　Elite and Specialized Interviewing. Evanston, Ill.: Northwest-
　　　　ern University Press.
DIAMANT, A.
　1966　The Nature of Political Development. *In* Political Development
　　　　and Social Change, J. Finkle and R. Gable, eds. New York:
　　　　Wiley.
DODD, L. C.
　1972　The Impact of Party System Characteristics on Cabinet Dura-
　　　　bility: A Game Theoretic Analysis. Paper presented to the
　　　　Annual Meeting of the Midwest Political Science Associa-
　　　　tion, Chicago.
DOGAN, M., and S. ROKKAN, eds.
　1969　Quantitative Ecological Analysis in the Social Sciences. Cam-
　　　　bridge: M.I.T. Press.
DOMHOFF, WILLIAM
　1967　Power in America. Englewood Cliffs, N.J.: Spectrum.
DONOVAN, JOHN C.
　1967　The Politics of Poverty. New York: Pegasus.
DOUGLAS, W.
　1963　The Role of Political Parties in the Modernization Process.
　　　　Koreana Quarterly 5:37–42.
DRAPER, N., and H. SMITH
　1967　Applied Regression Analysis. New York: Wiley.
DUHEM, PIERRE
　1962　The Aim and Structure of Physical Theory. Trans. Philip P.
　　　　Wiener. New York: Atheneum.
DUNCAN, OTIS DUDLEY
　1966　Path Analysis: Sociological Examples. American Journal of
　　　　Sociology 72:1–16. Reprinted in Causal Models in the Social
　　　　Sciences, H. M. Blalock, Jr., ed. Chicago: Aldine-Atherton,
　　　　1971.

　1969a　Some Linear Models for Two-Wave, Two-Variable Panel Analy-
　　　　sis, with One-Way Causation and Measurement Error. Psy-
　　　　chological Bulletin 72:177–82. *In* Mathematics and Sociology,
　　　　H. M. Blalock, Jr., et al., eds., forthcoming.

1969b   Contingencies in Constructing Causal Models. *In* Sociological Methodology 1969, E. F. Borgatta, ed. San Francisco, Calif.: Jossey-Bass.

DURKHEIM, EMILE

1933   The Division of Labor in Society. Trans. George Simpson. New York: Free Press.

1951   Suicide. Trans. George Simpson. New York: Free Press.

DYE, THOMAS R.

1972   Understanding Public Policy. Englewood Cliffs, N.J.: Prentice-Hall.

EASTON, D.

1969   The New Revolution in Political Science. American Political Science Review 63:1051–61.

ECKENRODE, R. T.

1965   Weighting Multiple Criteria. Management Science 12:180–92.

EISENSTADT, SHUMEL NOAH

1962–   Initial Institutional Patterns of Political Modernization.

1963   Civilizations (Brussels) 12:461–72; 13:15–26. Reprinted in Political Modernization: A Reader in Comparative Political Change, C. Welch, ed. Belmont, Calif.: Wadsworth, 1967.

1965   Essays on Comparative Institutions. New York: Wiley.

1967   Political Systems of Empires. New York: Free Press.

ERBE, WILLIAM

1964   Social Involvement and Political Activity: A Replication and Elaboration. American Sociological Review 29:198–215.

ERHARD, LUDWIG

1964   Speech cited in Germany in Europe. New York: German Information Center.

ERIKSON, R.

1971   The Electoral Impact of Congressional Roll Call Voting. American Political Science Review 65:1018–32.

ETZIONI, A.

1968   "Shortcuts" to Social Change? The Public Interest 12:40–51.

ETZIONI, A., and E. LEHMAN

1967   Some Dangers in "Valid" Social Measurement. Annals of the American Academy of Political and Social Science 373:1–15.

EVAN, W. M.

1959   Cohort Analysis of Survey Data: A Procedure for Studying Long-Term Opinion Change. Public Opinion Quarterly 23:63–72.

EVANS, J., and J. SCHILLER

1970   How Preoccupation with Possible Regression Artifacts Can Lead to a Faulty Strategy for the Evaluation of Social Action Programs: A Reply to Campbell and Erlebacher. *In* Compensatory Education: A National Debate, J. Hellmuth, ed. (Vol. 3 of The Disadvantaged Child). New York: Brunner/Mazel.

FELDMAN, HERMAN
1937   Problems in Labor Relations: A Case Book Presenting Some Major Issues in Relations of Labor, Capital and Government. New York: Macmillan.

FESTINGER, LEON
1965   Laboratory Experiments. *In* Research Methods in the Behavioral Sciences, Leon Festinger and Daniel Katz, eds. New York: Holt, Rinehart and Winston.

FISCHER, J. L.
1971   The Uses of Internal Revenue Service Income Information for Measuring the Impact of Manpower Programs. *In* A Conference on the Evaluation of the Impact of Manpower Programs, M. E. Borus, ed. Columbus, Ohio.

FOLTZ, W. J.
1963   Building the Newest Nations: Short-Run Strategies and Long-Run Problems. *In* Nation-Building, K. Deutsch and W. J. Foltz, eds. New York: Atherton.

FOX, K.
1968   Intermediate Economic Statistics. New York: Wiley.

FREEMAN, A. MYRICK, III
1970   Project Design and Evaluation with Multiple Objectives. *In* Public Expenditures and Public Analysis, R. H. Haveman and J. Margolis, eds. Chicago: Markham.

FRENCH, JOHN R. P., JR.
1965   Experiments in Field Settings. *In* Research Methods in the Behavioral Sciences, Leon Festinger and Donald Katz, eds. New York: Holt, Rinehart and Winston.

GALTUNG, J.
1964   A Structural Theory of Aggression. Journal of Peace Research 2:95–119.

1967   Theory and Methods of Social Research. Oslo: Universitetsforloget; London: Allen and Unwin; New York: Columbia University Press.

GERTH, H., and C. WRIGHT MILLS
1958   From Max Weber. New York: Oxford University Press.

GILLESPIE, J. V., and B. A. NESVOLD
1971   Macro Quantitative Analysis. Beverly Hills, Calif.: Sage.

GLASER, WILLIAM A.
1959   The Family and Voting Turnout. Public Opinion Quarterly 23:563–70.

GLASS, GENE V.
1968   Analysis of Data on the Connecticut Crackdown as Time-Series Quasi-Experiment. Law and Society Review 3:55–76.

GLASS, GENE V., G. C. TIAO, and T. O. MAGUIRE
1971   Analysis of Data on the 1900 Revision of German Divorce Laws as a Time-Series Quasi-Experiment. Law and Society Review 6:539–62.

GOLDBERG, A. S.
1966   Discerning a Causal Pattern among Data on Voting Behavior. American Political Science Review 60:913–22. Reprinted in Causal Models in the Social Sciences, H. M. Blalock, Jr., ed. Chicago: Aldine-Atherton, 1971.

GOLEMBIEWSKI, R., and A. BLUMBERG
1967   Confrontation as a Training Design in Complex Organizations: Attitudinal Changes in a Diversified Population of Managers. Journal of Applied Behavioral Science 3:525–47.

GOVE, WALTER, and HERBERT COSTNER
1969   Organizing the Poor: An Evaluation of a Strategy. Social Science Quarterly 50:643–54.

GRAHAM, H., and T. GURR, eds.
1969   Conclusion. Violence in America. New York: Signet.

GREENWOOD, E.
1945   Experimental Sociology: A Study in Method. New York: King's Crown.

GREER, SCOTT, and PETER ORLEANS
1962   The Mass Society and Parapolitical Structures. American Sociological Review 27:634–46.

GROSS, B. M.
1966   The State of the Nation: Social System Accounting. London: Tavistock. Also in Social Indicators, R. M. Bauer, ed. Cambridge: M.I.T. Press, 1966.

GROSS, B. M., ed.
1967   Social Goals and Indicators. Annals of the American Academy of Political and Social Science 371 (pt. 1, May):i–iii, 1–177; (pt. 2, September):i–iii, 1–218.

GUETZKOW, HAROLD
1968   Some Correspondences between Simulations and Realities. *In* New Approaches to International Relations, Morton Kaplan, ed. New York: St. Martin's.

GURIN, PATRICIA, GERALD GURIN, C. ROSINA, and MURIAL BEATTIE
1969   Internal-External Control in the Motivational Dynamics of Negro Youth. Journal of Social Issues 25:29–53.

GURR, TED ROBERT
1970   Why Men Rebel. Princeton, N.J.: Princeton University Press.
1971   Violence, Political Revolution and Social Change. Paper prepared for a seminar on Revolution and Social Change, Pennsylvania State University.

GUTTMAN, L.
1946   An Approach for Quantifying Paired Comparisons and Rank
        Order. Annals of Mathematical Statistics 17:144–63.

HAAS, ERNST B.
1964   Beyond the Nation-State: Functionalism and International
        Integration. Stanford, Calif.: Stanford University Press.

HAAS, MICHAEL, and THEODORE L. BECKER
1970   The Behavioral Revolution and After. In Approaches to the
        Study of Political Science, Michael Haas and Henry S. Kariel,
        eds. Scranton, Pa.: Chandler.

HALL, D. T., and B. MANSFIELD
1971   Organizational and Individual Response to External Stress.
        Administrative Science Quarterly 16:533–47.

HARMAN, HARRY H.
1967   Modern Factor Analysis. Chicago: University of Chicago Press.

HAUSER, R. M., and A. S. GOLDBERGER
1971   The Treatment of Unobservable Variables in Path Analysis.
        In Sociological Methodology 1971, H. L. Costner, ed. San
        Francisco: Jossey-Bass.

HAUSKNECHT, MURRAY
1962   The Joiners – A Sociological Description of Voluntary Associa-
        tion Membership in the United States. New York: Bedminster.

HAVEMAN, ROBERT H.
1972   The Economic Performance of Public Investments. Baltimore,
        Md.: Johns Hopkins Press.

HAYS, W.
1963   Statistics for Psychologists. New York: Holt.

HAZLEWOOD, LEO
1973   Externalizing Systemic Stresses, International Conflict as
        Adaptive Behavior. In Conflict Behavior and Linkage Politics,
        J. Wilkenfeld, ed. New York: David McKay.

HEISE, D. R.
1969   Separating Reliability and Stability in Test-Retest Correla-
        tions. American Sociological Review 34:93–101. Reprinted in
        Causal Models in the Social Sciences, H. M. Blalock, Jr., ed.
        Chicago: Aldine-Atherton, 1971.

1970   Causal Inference from Panel Data. In Sociological Method-
        ology 1970, E. Borgatta and G. W. Bohrnstedt, eds. San Fran-
        cisco: Jossey-Bass.

HELLER, R. N.
1971   The Uses of Social Security Administration Data for Measuring
        the Impact of Manpower Programs. In A Conference on the
        Evaluation of the Impact of Manpower Programs, M. E. Borus,
        ed. Columbus, Ohio.

HELLMUTH, J., ed.
1970   Compensatory Education: A National Debate (Vol. 3 of the Disadvantaged Child). New York: Brunner/Mazel.

HESS, R. L., and LOEWENBERG, G.
1964   The Ethiopian No-Party State: A Note on the Functions of Political Parties in Developing States. American Political Science Review 58:947-50.

HESSE, MARY
1967   Laws and Theories. *In* The Encyclopedia of Philosophy, 4. New York: Macmillan and The Free Press.

HIGHAM, JOHN
1967   Strangers in the Land. New York: Atheneum.

HILTON, GORDON
1972   Causal Inference Analysis: A Seductive Process. Administrative Science Quarterly 17:44-54.

HOBSBAWN, F. J.
1969   Industry and Empire. Middlesex, England: Pelican.

HOFFMAN, STANLEY
1965   The State of War. New York: Praeger.

HOFSTADTER, RICHARD
1970   Reflections on Violence in the United States. *In* American Violence, Richard Hofstadter and Michael Wallace, eds. New York: Random House.

HOFSTADTER, RICHARD, and MICHAEL WALLACE
1970   American Violence. New York: Random House.

HOPKINS, R.
1969   Political Roles and Political Institutionalization: The Tanzanian Experience. Paper delivered at the American Political Science Association convention, New York.

HOVLAND, CARL I.
1959   Reconciling Conflicting Results Derived from Experimental and Survey Studies. American Psychologist 14:8-17.

HOWARD, KENNETH L., and MERTON S. KRAUSE
1970   Some Comments on "Techniques for Estimating the Source and Direction of Influence in Panel Data." Psychological Bulletin 74:219-24.

HUBER, G. P., and A. DELBECQ
1971   Guidelines for Combining the Judgments of Individual Members in Decision Conferences. Academy of Management Journal 15:161-74.

HULL, CLARK L.
1950   A Primary Social Science Law. Scientific Monthly 71:221-28.

HUNTINGTON, S. P.
1965   Political Development and Political Decay. World Politics 17:386-430.

1968 Political Order in Changing Societies. New Haven, Conn.: Yale University Press.

HYMAN, H. H.
1972 Secondary Analysis of Sample Surveys: Principles, Procedures and Potentialities. New York: Wiley.

HYMAN, H. H., and C. R. WRIGHT
1967 Evaluating Social Action Programs. *In* The Uses of Sociology, P. F. Lazarsfeld, W. H. Sewell, and H. L. Wilensky, eds. New York: Basic Books.

INGLEHART, RONALD
1967 An End to European Integration? American Political Science Review 61:91–105.

1971 The Silent Revolution in Europe: Intergenerational Change in Post-Industrial Societies. American Political Science Review 65:991–1017.

INSTITUTE OF SCRAP IRON AND STEEL, INC.
1970 Facts Yearbook 1970, 31st ed. Washington, D.C.: Institute of Scrap Iron and Steel, Inc.

JACKSON, ELTON, and RICHARD CURTIS
1968 Conceptualization and Measurement in the Study of Social Stratification. *In* Methodology in Social Research, H. Blalock and A. Blalock, eds. New York: McGraw-Hill.

JACOB, HERBERT
1972 Contact with Government Agencies: A Preliminary Analysis of the Distribution of Government Services. Midwest Journal of Political Science 16:123–46.

JACOB, HERBERT, and MICHAEL LIPSKY
1968 Outputs, Structure and Power: An Assessment of Change in the Study of State and Local Politics. Journal of Politics 30:510–38.

JANDA, KENNETH
1971 A Technique for Assessing the Conceptual Equivalence of Institutional Variables across and within Culture Areas. Paper delivered at the American Political Science convention, Chicago.

JENKINS, ROBIN
1969 Ethnic Conflict and Class Consciousness: A Case Study from Belgium. *In* IPRA Studies on Peace Research Third Conference 3:122–37. Assen, Netherlands: Van Gorcum.

JOHNSTON, J.
1963 Econometric Methods. 2d ed., 1972. New York: McGraw-Hill.

KAHNEMAN, D.
1965 Control of Spurious Association and the Reliability of the Controlled Variable. Psychological Bulletin 64:326–29.

KAMISAR, Y.
1964   The Tactics of Police Persecution-Oriented Critics of the Courts. Cornell Law Quarterly 49:458–71.

KAPLAN, ABRAHAM
1964   The Conduct of Inquiry. San Francisco, Calif.: Chandler.

KAPLAN, MORTON A.
1957   System and Process in International Politics. New York: Wiley.

KATZ, DANIEL, and ROBERT KAHN
1966   The Social Psychology of Organizations. New York: Wiley.

KAYSEN, C.
1967   Data Banks and Dossiers. The Public Interest 7:52–60.

KENNY, DAVID A.
1970   Common Factor Model with Temporal Erosion for Panel Data. Paper presented at Social Science Research Council Conference on Structural Equation Models, Madison, Wisconsin.

1973   Cross-Lagged and Synchronous Common Factors in Panel Data. *In* Structural Equation Models in the Social Sciences, A. S. Goldberger and O. D. Duncan, eds. New York: Academic.

KEOHANE, R. O.
1969   Institutionalization in the United Nations General Assembly. International Organization 23:859–96.

KERLINGER, FRED N.
1964   Foundations of Behavioral Research. New York: Holt, Rinehart and Winston.

KEY, V. O.
1964   Public Opinion and American Democracy. New York: Knopf.

KNORR, K., and J. ROSENAU
1969   Contending Approaches to International Politics. Princeton, N.J.: Princeton University Press.

KNUPFER, GENEVIEVE
1947   Portrait of the Underdog. Public Opinion Quarterly 11:103–14.

KOLKO, GABRIEL
1969   The Roots of American Foreign Policy. Boston: Beacon.

KORMAROVSKY, MIRRA
1946   The Voluntary Associations of Urban Dwellers. American Sociological Review 11:686–98.

KORNHAUSER, WILLIAM
1959   The Politics of Mass Society. Glencoe, Ill.: Free Press.

LAND, KENNETH C.
1969   Principles of Path Analysis. *In* Sociological Methodology 1969, E. F. Borgatta, ed. San Francisco, Calif.: Jossey-Bass.

1971   "Significant Others, the Self-Reflexive Act and the Attitude Formation Process": A Reinterpretation. American Sociological Review 36:1085–98.

LaPonce, J. A.
   1970   Experimentation and Political Science: A Plea for More Pre-
          Data Experiments. Paper prepared for the International Politi-
          cal Science Association, Vancouver, Canada.
Lerner, D., and W. Schramm, eds.
   1967   Communication and Change in the Developing Countries.
          Honolulu, Hawaii: East-West Center Press.
Levenson, B., and M. S. McDill
   1966   Vocational Graduates in Auto Mechanics: A Followup Study
          of Negro and White Youth. Phylon 27:347–57.
Levy, Sheldon
   1969   A 150 Year Study of Political Violence in the United States.
          In Violence in America, H. Graham and T. Gurr, eds. New
          York: Signet.
Lieberman, S.
   1956   The Effects of Changes in Roles on the Attitudes of Role Occu-
          pants. Human Relations 9:385–402.
Lijphart, A.
   1971   Comparative Politics and Comparative Methoa. American
          Political Science Review 65:682–93.
Lindberg, Leon
   1971   Interest Groups in the EEC. In Regional International Or-
          ganizations: Structures and Functions, Paul A. Tharp, Jr., ed.
          New York: St. Martin's.
Lindberg, Leon, and Stuart A. Scheingold
   1970   Europe's Would-Be Policy. Englewood Cliffs, N.J.: Prentice-
          Hall.
Link, Arthur
   1966   American Epoch: A History of the United States Since the
          1890's. 2d ed. New York: Knopf.
Lipset, Seymour Martin, Martin A. Trow, and James Coleman
   1956   Union Democracy. Glencoe, Ill.: Free Press.
Lord, F. M.
   1958   Further Problems in the Measurement of Growth. Educational
          and Psychological Measurement 18:437–54.
McClosky, Herbert
   1964   Consensus and Ideology in American Politics. American Po-
          litical Science Review 58:361–82.
McClosky, Herbert, and Harold E. Dahlgren
   1959   Primary Group Influence in Party Loyalty. American Political
          Science Review 53:757–76.
McClosky, Herbert, and John H. Schaar
   1965   Psychological Dimensions of Anomy. American Sociological
          Review 30:14–40.

McCormick, D.
1970   A Field Theory of Dynamic International Processes. Dimensionality of Nations Project, Research Report No. 30, University of Hawaii.

Mack, R.
1971   Planning on Uncertainty: Decision Making in Business and Government Administration. New York: Wiley.

McNemar, Quinn
1969   Psychological Statistics. New York: Wiley.

Macrae, Norman
1970   How the EEC Makes Decisions. Atlantic Community Quarterly 8:363–71.

Manniche, E., and D. P. Hayes
1957   Respondent Anonymity and Data Matching. Public Opinion Quarterly 21:384–88.

Mar, B. W.
1971   A System of Waste Discharge for the Management of Water Quality. Water Resources Research 7:1079–86.

March, James
1953   Husband-Wife Interaction Over Political Issues. Public Opinion Quarterly 17:461–70.

Mason, R. D.
1967   Statistical Techniques in Business and Economics. Homewood, Ill.: Irwin.

Meehl, Paul E.
1970   Nuisance Variables and the Ex Post Facto Design. *In* Analyses of Theories and Methods of Physics and Psychology, Michael Radner and Stephen Winokur, eds. Vol. 4 of the Minnesota Studies in the Philosophy of Science. Minneapolis: University of Minnesota Press.

Merewitz, Leonard, and Stephen H. Sosnick
1971   The Budget's New Clothes. Chicago: Markham.

Michels, Robert
1949   Political Parties. Glencoe, Ill.: Free Press.

Miller, James G.
1971   The Nature of Living Systems. Behavioral Science 16:277–301.

Miller, Warren E., and Donald E. Stokes
1963   Constituency Influence in Congress. American Political Science Review 57:45–56.

MIT Study Group
1967   The Transitional Process. *In* Political Modernization: A Reader in Comparative Political Change, C. Welch, ed. Belmont, Calif.: Wadsworth.

MORRISON, DONALD G., R. MITCHELL, J. PADEN, and H. M. STEVENSON
1972   Black Africa: A Handbook. New York: Free Press.

MORRISON, DONALD G., and HUGH MICHAEL STEVENSON
1972   Integration and Instability: Patterns of African Political Development. American Political Science Review 66:902–27.

MOSTELLER, FREDERICK
1968   Association and Estimation in Contingency Tables. Journal of the American Statistical Association 63:1–28.

MOSTELLER, FREDERICK, and DANIEL P. MOYNIHAN, eds.
1972   On Equality of Educational Opportunity. New York: Random House.

MOYNIHAN, DANIEL P.
1969   Maximum Feasible Misunderstanding. New York: Free Press.

MUELLER, JOHN F., ed.
1969   Approaches to Measurement in International Relations. New York: Appleton-Century-Crofts.

NAGEL, ERNEST
1961   The Structure of Science. New York: Harcourt, Brace and World.

NARDIN, TERRY
1971   Theories of Conflict Management. Peace Research Review 4:1–93.

NAROLL, RAOUL
1968   Some Thoughts on Comparative Method in Cultural Anthropology. In Methodology in Social Research, Hubert M. Blalock and Ann Blalock, eds. New York: McGraw-Hill.

NEAL, ARTHUR, and MELVIN SEEMAN
1964   Organization and Powerlessness: A Test of the Mediation Hypothesis. American Sociological Review 29:216–26.

NEWHOUSE, JOHN
1967   Collision in Brussels: The Common Market Crisis in June 1965. New York: Norton.

NUNNALLY, J.
1967   Psychometric Theory. New York: McGraw-Hill.

OFFICE OF CHILD DEVELOPMENT
1967   Head Start: A Manual on Policies and Instruction. Washington, D.C.: Government Printing Office.
1970   Transmittal Notice – Head Start Policy Manual, 70.2 Washington, D.C.: Government Printing Office.

OFFICE OF ECONOMIC OPPORTUNITY
n.d.   Project Head Start: Parent Involvement. Washington, D.C.: Government Printing Office.

OLSON, MANCUR, JR.
1963   Rapid Growth as a Destabilizing Force. Journal of Economic History 23:529–52.

OLSON, MANCUR, JR., and RICHARD ZECKHAUSER
1970   An Economic Theory of Alliances. *In* Alliance in International Politics, Julian R. Friedman, Christopher Bladen, and Steve Rosen, eds. Boston: Allyn and Bacon.

PARK, T. W.
1972   The Role of Distance in International Relations: A New Look at the Social Field Theory. Behavioral Science 17:337–48.

PARSONS, TALCOTT
1951   The Social System. London: Collier-Macmillan.

1953   A Revised Analytical Approach to the Theory of Social Stratification. *In* Class, Status, and Power, R. Bendix and S. M. Lipset, eds. Glencoe, Ill.: Free Press.

PELZ, D. C. and F. M. ANDREWS
1964   Detecting Causal Priorities in Panel Study Data. American Sociological Review 29:836–48.

PELZ, D. C., and R. A. LEW
1970   Heise's Causal Model Applied. *In* Sociological Methodology 1970, E. Borgatta and G. W. Bohrnstedt, eds. San Francisco, Calif.: Jossey-Bass.

PETERS, C. C., and W. R. VAN WOORHIS
1940   Statistical Procedures and Their Mathematical Bases. New York: McGraw-Hill.

PETTIGREW, T. F. et al.
1973   Busing: A Review of "The Evidence." The Public Interest 30:88–118.

PINARD, MAURICE
1968   Mass Society and Political Movements. American Journal of Sociology 73:682–90.

PINDER, JOHN, and ROY PRYCE
1969   Europe after DeGaulle. Middlesex, England: Penguin.

POLANYI, M.
1966   The Tacit Dimension. New York: Doubleday.

1967   The Growth of Science in Society. Minerva 5:533–45.

POLSBY, N.
1968   The Institutionalization of the U.S. House of Representatives. American Political Science Review 62:144–68.

POMPER, G. M.
1972   From Confusion to Clarity: Issues and American Voters, 1956–1968. American Political Science Review 66:415–28.

POOL, ITHIEL DE SOLA, ROBERT P. ABELSON, and SAMUEL L. POPKIN
1964   Candidates, Issues and Strategies: A Computer Simulation of the 1960 Presidential Election. Cambridge: M.I.T. Press.

POPPER, F.
1971   Internal War as a Stimulant to Political Development. Comparative Political Studies 3:413–23.

POPPER, K. R.
  1959   The Logic of Scientific Discovery. New York: Basic Books.
  1963   Conjectures and Refutations. London: Routledge and Kegan
         Paul; New York: Basic Books.

PRZEWORSKI, ADAM, and HENRY TEUNE
  1970   The Logic of Comparative Social Inquiry. New York: Wiley.

PUCHALA, DONALD
  1970   International Transactions and Regional Integration. Inter-
         national Organization 24:732–63.

PUGH, D. S., and D. J. HICKSON
  1972   Causal Inference and the Aston Studies. Administrative
         Science Quarterly 17:273–76.

PYE, L.
  1965   The Concept of Political Development. Princeton, N.J.: Prince-
         ton University Press.

QUENOUILLE, MAURICE HENRY
  1952   Associated Measurements. New York: Academic.

RAPPAPORT, A.
  1964   Strategy and Conscience. New York: Schocken.

RASER, J. R.
  1969   Simulation and Society. Boston: Allyn and Bacon.

REUTERS INTERNATIONAL NEWS AGENCY
  1967   The New Africans: A Guide to the Contemporary History of
         Emergent Africa and Its Leaders. New York: Putnam.

RHEINSTEIN, M.
  1959   Divorce and the Law in Germany: A Review. American Journal
         of Sociology 65:489–98.

ROBINSON, JOHN P., JERROLD G. RUSK, and KENDRA B. HEAD
  1968   Measures of Political Attitudes. Ann Arbor, Mich.: Survey
         Research Center, Institute for Social Research.

ROGOWSKI, RONALD L. and LOIS WASSERSPRING
  1971   Does Political Development Exist? Corporatism in Old and
         New Societies. Beverly Hills, Calif.: Sage.

ROOS, LESLIE L., JR., and NORALOU P. ROOS
  1971   Managers of Modernization. Cambridge: Harvard University
         Press.
  1972   Pollution, Regulation and Evaluation. Law and Society Review
         6:509–29.

ROSE, A. M.
  1952   Needed Research on the Mediation of Labor Disputes. Per-
         sonnel Psychology 5:187–200.

ROSE, STEPHEN M.
  1972   The Betrayal of the Poor. Cambridge: Schenkman.

ROSENAU, JAMES
1969   Linkage Politics. New York: Free Press.

ROSENBERG, MORRIS
1951   The Meaning of Politics in Mass Society. Public Opinion Quarterly 15:5–15.

1954   Some Determinants of Political Apathy. Public Opinion Quarterly 18:349–66.

1965   Society and the Adolescent Self-Image. Princeton, N.J.: Princeton University Press.

Ross, H. L., and D. T. CAMPBELL
1968   The Connecticut Speed Crackdown: A Study of the Effects of Legal Change. *In* Perspectives on the Social Order: Readings in Sociology, H. L. Ross, ed. New York: McGraw-Hill.

Ross, H. L., D. T. CAMPBELL, and G. V. GLASS
1970   Determining the Social Effects of a Legal Reform: The British "Breathalyser" Crackdown of 1967. American Behavioral Scientist 13:493–509.

ROZELLE, R. M., and D. T. CAMPBELL
1969   More Plausible Rival Hypotheses in the Cross-Lagged Panel Correlation Technique. Psychological Bulletin 71:74–80.

RUBENSTEIN, A. H., M. RADNOR, N. R. BAKER, D. R. HEIMAN, and J. McCOLLY
1967   Some Organizational Factors Related to the Effectiveness of Management Science Groups in Industry. Management Science 18:B508–B518.

RUMMEL, RUDOLPH J.
1963   Dimensions of Conflict Behavior Within and Between Nations. General Systems Yearbook 8:1–50.

1971a   Applied Factor Analysis. Evanston, Ill.: Northwestern University Press.

1971b   A Status-Field Theory of International Relations. Dimensionality of Nations Project, Research Report No. 50, University of Hawaii.

SAWYER, J., and H. SCHECHTER
1968   Computers, Privacy, and the National Data Center: The Responsibility of Social Scientists. American Psychologist 23:810–18.

SCHANCK, R. L., and C. GOODMAN
1939   Reactions to Propaganda on Both Sides of a Controversial Issue. Public Opinion Quarterly 3:107–12.

SCHATTSCHNEIDER, E. E.
1960   The Semi Sovereign People. New York: Holt, Rinehart and Winston.

SCHEUCH, ERWIN K.
1966   Cross-National Comparisons Using Aggregate Data: Some Substantive and Methodological Problems. *In* Comparing Nations: The Use of Quantitative Data in Cross-National Research, R. L. Merritt and Stein Rokkan, eds. New Haven, Conn.: Yale University Press.

SCHMITTER, PHILIPPE C.
1970   Central American Integration: Spillover, Spill-Around or Encapsulation? Journal of Common Market Studies 9:1-48.

SCHRAMM, W.
1964   Mass Media and National Development. Stanford, Calif.: Stanford University Press.

SCHULTZE, CHARLES L., EDWARD R. FRIED, ALICE M. RIVLIN, and NANCY H. TEETERS
1972   Setting National Priorities: The 1973 Budget. Washington, D.C.: Brookings Institution.

SCHWARTZ, R. D.
1961   Field Experimentation in Sociological Research. Journal of Legal Education 13:401-10.

SCHWARTZ, R. D., and S. ORLEANS
1967   On Legal Sanctions. University of Chicago Law Review 34:274-300.

SCHWARTZ, R. D., and J. H. SKOLNIC
1963   Televised Communication and Income Tax Compliance. *In* Television and Human Behavior, L. Arons and M. Mays, eds. New York: Appleton-Century-Crofts.

SCROLE, LEO J.
1956   Social Integration and Certain Corollaries: An Exploratory Study. American Sociological Review 21:709-16.

SEEMAN, MELVIN
1959   On the Meaning of Alienation. American Sociological Review 24:783-91.

1966   Alienation, Membership, and Political Knowledge: A Comparative Study. Public Opinion Quarterly 30:353-67.

1972   Alienation and Engagement. *In* The Human Meaning of Social Change, Angus Campbell and Philip E. Converse, eds. New York: Russell Sage Foundation.

SEGAL, RONALD
1961   Political Africa: A Who's Who of Personalities and Parties. London: Stevens.

SELLTIZ, CLAIRE, MARIE JAHODA, MORTON DEUTSCH, and STUART W. COOK
1966   Research Methods in Social Relations. New York: Holt, Rinehart and Winston.

SELVIN, H.
   1957   A Critique of Tests of Significance in Survey Research. American Sociological Review 22:519–27.
SHARKANSKY, IRA, and RICHARD I. HOFFERBERT
   1969   Dimensions of State Politics, Economics and Public Policy. American Political Science Review 63:867–79.
SILVERT, K.
   1965   Parties and Masses. Annals of the American Academy of Political and Social Science 358:101–8.
SIMON, H. A.
   1971   Spurious Correlation, A Causal Approach. *In* Causal Models in the Social Sciences, H. Blalock, Jr., ed. Chicago: Aldine-Atherton.
SIMON, J. L.
   1966   The Price Elasticity of Liquor in the U.S. and a Simple Method of Determination. Econometrica 34:193–205.
SINGER, J. DAVID
   1971   A General Systems Taxonomy for Political Science. New York: General Learning Press.
SKINNER, B. F.
   1969   Contingencies of Reinforcement: A Theoretical Analysis. New York: Appleton-Century-Crofts.
SMOKER, PAUL
   1969   A Time Series Analysis of Sino-Indian Relations. Journal of Conflict Resolution 13:172–91.
SOLOMON, R. W.
   1949   An Extension of Control Group Design. Psychological Bulletin 46:137–50.
SOROKIN, P.
   1937   Social and Political Dynamics. Vol. 3 of Fluctuations of Social Relationships, War and Revolution. New York: American Book.
SPINELLI, ALTIERO
   1966   The Eurocrats. Baltimore, Md.: Johns Hopkins Press.
STANLEY, JULIAN C.
   1967   On Improving Certain Aspects of Educational Experimentation. *In* Improving Experimental Design and Statistical Analysis, Julian Stanley, ed. Chicago: Rand McNally.
STEVENSON, H. M.
   1971   Conflict and Instability in Africa. Ph.D. dissertation, Northwestern University.
STIEBER, J. W.
   1949   Ten Years of the Minnesota Labor Relations Act. Minneapolis, Minn.: Industrial Relations Center, University of Minnesota.

STIMSON, D. H., and R. H. STIMSON
1971   Operations Research and Systems Analysis in Hospital Administration. Working Paper 147, Institute of Urban and Regional Development, University of California, Berkeley.

STOHL, MICHAEL
1971   The Study of Conflict Behavior Within and Between Nations: Some New Evidence. Paper presented to the Annual Meeting of the Midwest Political Science Association, Chicago.

STOUFFER, S. A.
1949   The Point System for Redeployment and Discharge. In Combat and Its Aftermath (Vol. 2 of The American Soldier), S. A. Stouffer et al., eds. Princeton, N.J.: Princeton University Press.

STROUSE, JAMES C., and J. OLIVER WILLIAMS
1972   A Non-Additive Model for State Policy Research. Journal of Politics 34:648-57.

SUCHMAN, E. A.
1967   Evaluative Research: Principles and Practice in Public Service and Social Action Programs. New York: Russell Sage Foundation.

SULLIVAN, J. L.
1971   Multiple Indicators and Complex Causal Models. In Causal Models in the Social Sciences, H. M. Blalock, Jr., ed. Chicago: Aldine-Atherton.

SUMMERS, G. F.
1970   Attitude Measurement. Chicago: Rand McNally.

SWEEN, JOYCE, and DONALD T. CAMPBELL
1965   A Study of the Effect of Proximally Autocorrelated Error on Tests of Significance for the Interrupted Time-Series Quasi-Experimental Design. Mimeographed. Department of Psychology, Northwestern University.

TAFT, PHILIP, and PHILIP ROSS
1969   American Labor Violence: Its Causes, Character and Outcome. In Violence in America, H. Graham and T. Gurr, eds. New York: Signet.

TANTER, RAYMOND
1964   Dimensions of Conflict Behavior Within and Between Nations 1958-1960. Ph.D. dissertation, Indiana University.

1965   Dimensions of Conflict Behavior Within Nations, 1955-60: Turmoil and Internal War. Peace Research Society Papers 3:159-84.

1966   Dimensions of Conflict Behavior Within and Between Nations, 1958-1960. Journal of Conflict Resolution 10:41-64.

TEUNE, HENRY
1971   Integration of Political Systems. Paper delivered at the Annual

Meeting of the International Studies Association, San Juan, Puerto Rico.

THISTLETHWAITE, DONALD L., and DONALD T. CAMPBELL
1960 Regression-Discontinuity Analysis: An Alternative to Ex Post Facto Experiment? Journal of Educational Psychology 51:309–17.

THOMPSON, C., and G. RATH
1972 The Administrative Experiment: A Special Case of Field Testing or Evaluation. Working Paper, Department of Industrial Engineering and Management Science, Northwestern University.

THOMPSON, WAYNE E., and JOHN E. HORTON
1962 Powerlessness and Political Negativism: A Study of Defeated Local Referendums. American Journal of Sociology 67:485–93.

THORNER, R. M.
1971 Health Program Evaluation in Relation to Health Programming. HSMHA Health Reports 86:525–32.

TILLY, CHARLES
1969 Collective Violence in European Perspective. *In* Violence in America, H. Graham and T. Gurr, eds. New York: Signet.

TORGERSON, WARREN
1958 Theory and Methods of Scaling. New York: Wiley.

VANECKO, JAMES J.
1969 Community Mobilization and Institutional Change: The Influence of the Community Action Program in Large Cities. Social Science Quarterly 50:609–30.

VINCENT, J. E.
1970 Factor Analysis in International Relations. Gainesville, Fla.: University of Florida Press.

VON NEUMANN, J., and O. MORGENSTERN
1947 Theory of Games and Economic Behavior. Princeton, N.J.: Princeton University Press.

WALKER, H. M., and J. LEV
1953 Statistical Inference. New York: Holt, Rinehart and Winston.

WALKER, JACK L.
1969 The Diffusion of Innovations among the American States. American Political Science Review 63:880–99.

WALLERSTEIN, I.
1966 The Decline of the Party in Single-Party African States. *In* Political Parties and Political Development, J. La Palombara and M. Weiner, eds. Princeton, N.J.: Princeton University Press.

WARNER, LLOYD, and PAUL LUNT
1941 The Social Life of a Modern Community. New Haven, Conn.: Yale University Press.

WEBB, E. J., D. T. CAMPBELL, R. D. SCHWARTZ, and L. B. SECHREST
 1966 Unobtrusive Measures: Nonreactive Research in the Social
  Sciences. Chicago: Rand McNally.

WEBER, MAX
 1947 The Theory of Social and Economic Organizations. New York:
  Oxford University Press.

WEINER, M.
 1965 Political Integration and Political Development. Annals of
  the American Academy of Political and Social Science 358:52–
  64.

WEISS, R. S., and M. REIN
 1970 The Evaluation of Broad-Aim Programs: Experimental Design,
  Its Difficulties, and an Alternative. Administrative Science
  Quarterly 15:97–109.

WELFLING, MARY
 1971 Political Institutionalization: The Development of a Concept
  and Its Empirical Application to African Party Systems. Ph.D.
  dissertation, Northwestern University.

WENNER, LETTIE M.
 1972 Enforcement of Water Pollution Control Laws. Law and Society
  Review 6:481–507.

WHITE, H.
 1970 Chains of Opportunity. Cambridge: Harvard University Press.

WHOLEY, J. S., J. W. SCALON, H. G. DUFFY, J. S. FUKUMOT, and L. M. VOGT
 1970 Federal Evaluation Policy. Washington, D.C.: Urban Institute.

WIGGINS, JAMES A.
 1968 Hypothesis Validity and Experimental Laboratory Methods.
  In Methodology in Social Research, Hubert M. Blalock, Jr.,
  and Ann B. Blalock, eds. New York: McGraw-Hill.

WILDAVSKY, A.
 1964 The Politics of the Budgetary Process. Boston: Little, Brown.

WILEY, D. F., and J. A. WILEY
 1970 The Estimation of Measurement Error in Panel Data. American Sociological Review 35:112–17. Reprinted in Causal Models
  in the Social Sciences, H. M. Blalock, Jr., ed. Chicago: Aldine-
  Atherton, 1971.

WILEY, NORBERT
 1969 America's Unique Class Politics: The Interplay of the Labor,
  Credit, and Commodity Markets. In Recent Sociology, Hans
  Peter Drietzel, ed. New York: Macmillan.

WILKENFELD, J.
 1968 Domestic and Foreign Conflict Behaviour of Nations. Journal
  of Peace Research 1:56–69.

 1969 Research Communication: Some Further Findings Researching the Domestic and Foreign Conflict Behaviour of Nations. Journal of Peace Research 2:147–55.

WILLIAMS, WALTER
1972　The Struggle for a Negative Income Tax. Public Policy Monograph No. 1, Institute of Governmental Research, University of Washington.

WILLIAMS, WALTER, and J. W. EVANS
1969　The Politics of Evaluation: The Case of Head Start. Annals of the American Academy of Political and Social Science 385:118–32.

WILLIAMS, WILLIAM A.
1970　The Roots of the Modern American Empire. New York: Vintage.

WILLNER, A. R., and D. WILLNER
1965　The Rise and Role of Charismatic Leaders. Annals of the American Academy of Political and Social Science 358:77–88.

WOLF, E., G. LÜKE, and H. HAX
1959　Scheidung und Scheidungsrecht: Grundfragen der Ehescheidung in Deutschland. Tübingen: Mohr.

WOLMAN, M. GORDON
1971　The Nation's Rivers. Science 174:905–18.

WRIGGINS, H.
1961　Impediments to Unity in New Nations: The Case of Ceylon. American Political Science Review 55:313–20.

ZINNES, DINA
1966　A Comparison of Hostile State Behavior in Simulate and Historical Data. World Politics 18:474–502.

ZOLBERG, A.
1963　Mass Parties and National Integration: The Case of the Ivory Coast. Journal of Politics 25:36–48.

ZURCHER, LOUIS A.
1969　Stages of Development in Poverty Program Neighborhood Action Committees. Journal of Applied Behavioral Science 5:222–67.

INDEX